ABSORPTION

SPECTROPHOTOMETRY

ABSORPTION SPECTROPHOTOMETRY

G. F. Lothian, M.A., F.Inst.P.

Senior Lecturer in Physics at the University of Exeter

THIRD EDITION

ADAM HILGER LTD
LONDON

PRINTED IN GREAT BRITAIN BY ROBERT MACLEHOSE AND CO. LTD, THE UNIVERSITY PRESS, GLASGOW

Preface to Third Edition

A new edition every ten years seems to be about right—1949, 1958, and now 1969. This third edition differs as much from the second as that did from the first. I am still pleased to acknowledge the use of some material from Twyman and Allsopp's *Practice of Absorption Spectrophotometry*. But while the design of the book remains unchanged, the 1958 edition has been ruthlessly revised. A lot of early work has been eliminated. The sections of part III dealing with instrument designs have been considerably reduced; it is impossible to be up to date as regards instrumentation, and in any case the proper place for the description and discussion of an instrument is the manufacturer's catalogue and ' Instructions for Use '. Only a very limited number of instruments have been mentioned to illustrate the various principles discussed; the inclusion or omission of any particular instrument should not be taken as an indication of the author's opinion on its merits. Some repetition has been avoided by mixing into one chapter descriptions of all types of spectrophotometer; the attention given to photographic and visual instruments has been considerably reduced— the latter almost to vanishing point.

Spectrofluorimetry has been introduced into this edition. It is a subject that is receiving increasing attention; and since fluorescence is a manifestation of absorption, it is very proper to include it in a volume with this title. Other new topics introduced are Fourier transform spectroscopy, the method of attenuated total reflection (ATR), and estimation of particle size by ' apparent ' absorption spectra, and the spectrophotometric determination of required paint and dye mixtures.

As regards symbols, recommendations of the International Unions of Pure and Applied Physics and Chemistry (IUPAP and IUPAC) have been used. The abbreviation for Ångstrom unit is Å, not A (A stands for ampere). The much used wavelength units micron (μ) and milli-micron (mμ) should now be replaced by micrometre (μm) and nano-metre (nm). These changes have been introduced in newly written sections. But some readers may be glad to find, because of lack of vigilance or the difficulty of remaking diagrams, occasional appearances of their old friends μ and mμ. The various systems of nomenclature (optical density, absorbance, etc.) have all been used so that readers new to the subject will become familiar with all the terms in common use.

It is hoped that in spite of the condensation that has been necessary in what is now a vast subject, sufficient literature references and sources for further reading are suggested to make the book a useful introduction.

The author is indebted to various reviewers and to others for drawing attention to errors and desirable changes. He is particularly grateful, for valuable comments and discussion, to Mr E. Bishop, Mr G. E. Fishter, Dr A. L. Glenn, Mr R. A. C. Isbell and Dr K. J. Ivin. In addition, the author is not a little indebted to Mr N. Goodman of Adam Hilger Ltd for a number of useful suggestions and for care in eliminating errors and obscurities. Acknowledgement for permission to reproduce diagrams is due to authors or to publishers and is indicated below the diagrams.

G.F.L.

Exeter, *August* 1968

Contents

PART I

The Principles of Spectrophotometry

CHAPTER I

Introduction

Early work

The earliest recorded observations of absorption spectra were made by Brewster in 1833, but the first serious work in the ultra-violet appears to have been that of Miller and Stokes. In 1862 they independently communicated to the Royal Society the results of experiments on the transparency of various substances in the ultra-violet. In Stokes' instrument, the ultra-violet rays were rendered visible by fluorescence, but Miller's instrument was a true quartz spectrograph. With it he determined the absorption of a number of substances. Of all the substances he tried, ice (and water), fluorspar and aqueous ammonia were the only ones that did not cut off some of the rays transmitted by the quartz spectrograph—although it would seem that other liquids (alcohol and glycerine) examined by him must have been impure or he would have found them as transparent as water. He also worked with gases and investigated the reflection from polished surfaces of many kinds—finding instances of absorption in both sets of experiments. In his paper he described the selective absorption of silver and the absorption spectra of twenty-one solids (eight of them glasses), 109 solutions, twenty-five gases, and sixteen polished reflecting surfaces.

In 1872 Hartley* came into possession of the instrument that Miller had used, improved it, and continued the earlier investigations. Miller had been unable to find any relation between constitution and absorption spectra, although he wrote (1864): 'The most interesting fact, however, disclosed by these various experiments is the persistence of either the diactinic or the absorbent property in the compound whatever its physical state— a circumstance which proves that the property under examination is intimately connected with the atomic or molecular nature of the body, and not merely with its state of aggregation.' We shall see that Hartley was to have rather more success.

* Hartley's instrument is now in the Science Museum, London.

In the hundred odd years since Brewster's observations, the spectra of very many substances have been measured. The information obtained is often so significant that the absorption spectrum of a newly produced compound is now one of the first properties to be investigated. Much of the work done before 1910 was unfortunately valueless because the methods employed were either qualitative or at most semi-quantitative. Hartley and later workers developed a method which recorded the wavelengths at which various thicknesses of a solution ceased to transmit. It was possible, by diluting the solution progressively, to follow the course of the absorption to maximum intensity and to plot a curve (equivalent thickness of standard solution against limiting wavelength transmitted) that produced the main characteristics of the absorption band.* The wavelengths of the maxima of absorption so recorded are fairly trustworthy, but the curves do not give any accurate idea of their intensity, and curves for two different substances cannot be compared unless the conditions of measurement were identical. The method also fails to reveal the detailed course of the absorption; see Fig. 1.1.

On the basis of extensive measurements of absorption spectra of this kind, Hartley, and a good deal later Baly and others, attempted to establish a relationship between the chemical constitution of an organic molecule and its absorption spectrum. They found that similarity in absorption spectrum often corresponds to similarity in structure, and they used this relationship to elucidate problems of molecular structure when ordinary chemical methods had failed. They were able to attribute the development of colour in a compound to the presence of certain ' chromophoric ' groupings known to be present in dyestuffs, and they also applied absorption measurements to the study of tautomerism, as in keto-enol equilibria. Modern work, however, has shown that applications of this type should only be made with great caution, and this will be more fully discussed later. For an account of this early work see Baly's *Spectroscopy* or Smiles' *The Relationship between Chemical Constitution and some Physical*

* For a description of the method see Kayser's *Handbuch der Spektroskopie*, Vol. III (Chapter 3 was written by Hartley) and Baly's *Spectroscopy*.

Properties or Kayser's *Handbuch der Spektroskopie*. Hartley must undoubtedly be considered the pioneer in this branch of the subject.

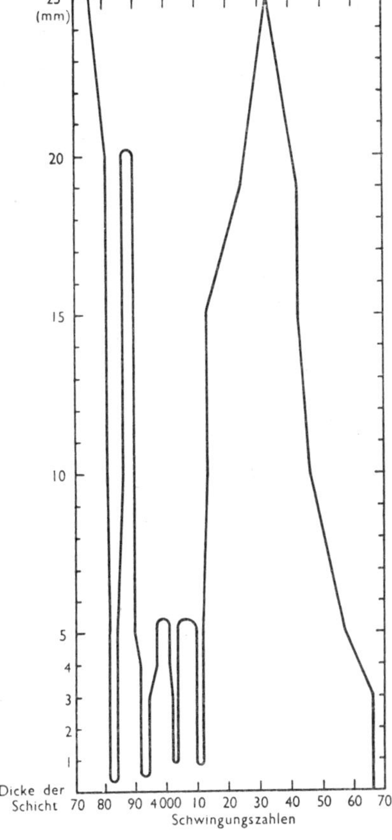

FIG. 1.1. Absorption spectrum of benzene (concentration 0·05 M) after Hartley. Abscissae represent wavenumbers and ordinates represent cell thicknesses in mm. (Adapted from Kayser's *Handbuch der Spektroskopie*.)

The investigations of Victor Henri

One of the disadvantages of these earlier measurements was that they were made chiefly with substances of very complex structure. The first systematic investigation of the ultra-violet absorption spectra of simple organic compounds was made by Victor Henri

(1919) and his pupils. Henri employed a photographic method that enabled him to plot molecular extinction coefficients against wavelengths λ.

The first result of Henri's investigations was to show that the absorption spectrum of a compound is not specifically characteristic of the whole molecule, but only of particular radicals in it. All simple ketones, for example, exhibit an absorption band with a maximum intensity about $\epsilon_{max} = 15$ between 2700 and 3000 Å; the absorption curve has the same characteristic shape for all ketones, but in a homologous series the maximum moves towards longer wavelengths and its intensity increases as the weight of the molecule increases. Other radicals behave in a similar way and Henri extended the use of the term ' chromophor ', which had previously been applied to the groups that could be associated with visible colour, to cover any radical that gives rise to such a characteristic absorption spectrum. When more than one chromophoric group is present in the molecule, the appearance of the absorption spectrum is changed.

The considerable progress that has been made in the empirical and theoretical treatment of chromophoric groups will be considered more fully in Chapter 6.

Terminology

Absorption. There are several alternative sets of names; those with some sort of official backing are summarized together with definitions in Table B (at back of book). They are more fully discussed in Chapter 3 on the laws of absorption. The names in column 1 were recommended in 1942 by the then Society of Public Analysts (now the Society for Analytical Chemistry). The other two systems originated in America and are slowly superseding the earlier British system. But all three sets of names are in common use, and a student of the subject must be familiar with them all. For this reason, all the names are used fairly indiscriminately in the present volume. A number of authors seem to have their own private nomenclatures, resulting sometimes in ambiguity of meaning, for which reason any departure from the terms and definitions of Table B is to be strongly deprecated.

Electromagnetic radiation. A wave may be specified by its wavelength λ, its frequency ν or wavenumber $\bar{\nu}$ (sometimes σ); wavenumber is the number of waves per unit length. An electromagnetic wave may be further specified by the energy E of one light quantum or photon. (Quantum theory is further discussed in Chapters 2 and 6.) Relations between the quantities are:

$$\bar{\nu} = 1/\lambda$$
$$\nu \times \lambda = v = c/n$$
$$E = h\nu$$

h is Planck's quantum of action
v is the velocity of a wave in a given medium
c is the velocity in vacuum
n is the refractive index of the medium.

The frequency ν is independent of the medium in which the radiation is travelling. But when monochromatic radiation leaves one medium and enters another, both the wavelength and the wavenumber change. In going from air to vacuum, the change amounts to about 3 parts in 10,000 ($n_{air} \approx 1 \cdot 0003$); the difference is sometimes significant. Values of wavelength normally quoted are for air. The following units are used:

Wavelength:

$$\text{Ångstrom (Å)} \quad = 10^{-10} \text{ m} = 10^{-8} \text{ cm}$$
$$\text{micrometre } (\mu\text{m}) \quad = 10^{-6} \text{ m} = 10^{-4} \text{ cm}$$
$$\text{nanometre (nm)} \quad = 10^{-9} \text{ m} = 10 \text{ Å}$$

In the past the names micron (μ) for 10^{-6} m and millimicron (mμ) for 10^{-9} m have been commonly used, and they are still being used; for this reason, μ and mμ have not been exhaustively removed from the parts of this book which appeared in earlier editions, but their future use is to be discouraged since the three listed above form part of an internationally agreed system.

Wavenumber:

This is almost universally expressed as the number of waves per cm, the unit being cm^{-1}. At one time (Bakker, 1952) it was recommended that this unit be named the Kayser (K), but the name has not come into general use.

Frequency:

The unit (for the optical region) is almost always vibrations per second—i.e. sec^{-1}. The unit is sometimes named the Hertz (Hz). Another unit is the fresnel—1 fresnel $= 10^{12}$ vibrations/sec. It is not usual for frequencies to be quoted for optical radiations.

The energy E of a photon may be expressed in ergs, joules, or calories. When the calorie is used, it is usual to quote the energy of N_A photons, where N_A is Avogadro's number, the number of molecules in a gram-molecule (mole). Thus, if one mole absorbs N_A photons to increase its energy by $N_A E$, its energy increase can be related to, and is possibly equal to, an energy of reaction that can be determined by thermochemical methods.

The electron-volt is a unit of energy equal to the energy acquired by an electron when it is accelerated through a potential difference of 1 volt. This is a particularly useful unit because, if an electron accelerated through V volts gives up all its energy on collision with an atom or molecule, the latter has enough energy to radiate a photon of frequency ν given by

$$Ve = h\nu = hc/\lambda$$

Insertion of the appropriate numerical values (see Table A, at back of book) into this expression leads to the useful expression

$$V \text{ (in volts)} \times \lambda \text{ (in Ångstroms)} = 12{,}398 \qquad (1.1)$$

Equivalent temperature

According to classical physics, at temperature T degrees absolute, one molecule has a mean energy $\frac{1}{2}kT$ per degree of freedom, where k is Boltzmann's constant. Such a molecule then has a reasonable chance of radiating a photon of frequency given by

$$h\nu = \tfrac{1}{2}kT$$

Typical values of the various modes of expressing an electromagnetic radiation are in Table C (back of book). From the column giving equivalent temperatures it will be appreciated that a molecule in thermal equilibrium with its surroundings in the laboratory has a reasonable chance of radiating a photon in the far infra-red; but that only at stellar temperatures can one expect any spontaneous radiation in the visible spectrum.

Spectral regions

A region of the spectrum may be described by referring to it as the visible region ($\lambda = 4000\text{–}7000$ Å and $\bar{\nu} = 25{,}000\text{–}14{,}000$ cm^{-1}), the ultra-violet ($\lambda < 4000$ Å and $\bar{\nu} > 25{,}000$ cm^{-1}), or the infra-red ($\lambda > 7000$ Å and $\bar{\nu} < 14{,}000$ cm^{-1}). Particular sections of the latter two regions may be specified by the type of prism material to which the radiations are transparent (see Chapter 5); thus the quartz region of the infra-red covers wavelengths to about 3 μm and the fluorite ultra-violet covers wavelengths down to about 1300 Å that cannot be reached with a quartz prism. The region covered by wavelengths less than about 1800 Å is sometimes spoken of as the vacuum ultra-violet because air absorbs these radiations and the apparatus has to be evacuated. The Schumann region of the ultra-violet covers wavelengths shorter than about 2000 Å; here the gelatine in normal photographic plates is absorbing and special plates, originally prepared by Schumann, have to be used.

CHAPTER 2

The Nature of Absorption.   Fluorescence

To understand what happens to a material when radiation is passed into it in order to measure its absorption spectrum, something must be known of quantum theory. This is an enormous subject, which can here be discussed but briefly. The theory is developed a little further in Chapter 6. Those who wish to pursue the subject are referred to the list of Further Reading at the end of the book.

Light quanta or photons

Measurements of black-body radiation and of the photoelectric effect were interpreted by Planck in 1901 and Einstein in 1905 by supposing that electromagnetic radiation is emitted and absorbed by matter, not continuously in arbitrary amounts, but in discrete fixed quantities, each having energy E, where

$$E = h\nu \tag{2.1}$$

ν is the frequency of the radiation and h is a constant known as Planck's quantum of action, or simply as Planck's constant. These ' packets ' of electromagnetic energy are named photons or light quanta. Furthermore, after emission from a substance, a photon does not spread out in all directions like a wave on water, but moves about as an entity and, in due course, is likely to be absorbed as a whole when it approaches matter. This seems to be contrary to the spreading waves which we think of in relation to the theory of, say, a diffraction grating. But we can consider the maxima and minima in a diffraction pattern to be positions of large and small probability for the arrival of photons; thus, a complete diffraction pattern is obtained only when we make observations involving a very large number of photons.

Quantized energy levels

With certain exceptions—some real, some only apparent— emission and absorption spectra are fundamentally line spectra.

This fact, combined with the photon concept, leads to the conclusion that the energy of matter can change only by discrete amounts and that only certain energy levels occur in the fundamental units of matter. To take a simple example, we know that hydrogen atoms emit, and under suitable conditions absorb, radiation of wavelength $\lambda = 6563$ Å (the Balmer red line). In accordance with the photon concept, such a photon must have energy $h\nu = hc/\lambda$, which (using Table A at the back of the book or expression 1.1 on page 8) equals $3 \cdot 0 \,.\, 10^{-19}$ joules or $1 \cdot 9$ electron-volts. This means that the hydrogen atom must have two energy levels E_1, E_2 differing by this amount, i.e.

$$h\nu = E_1 - E_2 \qquad (2.2)$$

More complete investigation of the hydrogen spectrum indicates that in fact there are no other energy levels of the hydrogen atom between these two.

Quantum theory of atoms and molecules

The first successful theory put forward to explain these quantized energy levels was that of Niels Bohr in 1913. Rutherford had shown, as a result of experiments on the scattering of α-particles, that an atom must have the whole of its positive charge concentrated in a small nucleus of diameter about 10^{-12} cm. Bohr proposed that the single electron present in a hydrogen atom was rotating around this nucleus, and that only certain orbits were permitted, namely those with

$$\text{angular momentum} = nh/2\pi \qquad (2.3)$$

Here, h is Planck's constant again, but this time it relates to matter rather than photons, and n is an integer named a quantum number. Further quantum numbers have since been introduced and n is now named the principal quantum number. Equation (2.3) was the central feature of this theory; it was quite arbitrary, its justification being that, combined with equation (2.2), it yielded correct values of the known wavelengths of the hydrogen spectrum and predicted the existence of wavelengths later discovered. The theory was thus highly successful in explaining at least the coarse structure of the spectrum of atomic hydrogen (and of ionized helium), but was of little help with any other spectra. The theory has now been superseded, but the picture of electron orbits is still a useful approximate concept. Very little further progress was made until, in 1924–5, de Broglie, Schrödinger and others developed wave mechanics.

Wave mechanics

We have already seen that electromagnetic radiation can sometimes be thought of as packets of energy and sometimes as a wave motion (diffraction, etc.). It was suggested that matter might also have a dual nature and that, while we normally think of matter as composed of particles, it might also have wave properties. This suggestion was confirmed experimentally in 1927 with the discovery of electron diffraction. We cannot here develop wave-mechanical theory (see Further Reading). It must suffice to state that the theory has developed into a very powerful method of showing the existence of quantized energy levels and that, in a number of cases, it is possible to calculate the positions of such energy levels. Some of the conclusions of wave mechanics that are relevant to the topics discussed in this volume are outlined below.

Electron quantum numbers

The energy levels of an electron in an atom or molecule can be characterized by four quantum numbers:

n, principal quantum number, similar to Bohr's one and only quantum number. Possible values are

$$n = 1, 2, 3, \ldots$$

l, orbital angular momentum quantum number, with possible values

$$l = 0, 1, 2, \ldots (n - l)$$

where

$$\text{angular momentum} = \sqrt{[l(l+1)]} \cdot h/2\pi \qquad (2.4)$$

A state of zero angular momentum is not visualized in the Bohr orbit picture; one can think of it in terms of a spherical distribution of charge.

m_l, magnetic quantum number, is the component of l along a preferred axis and thus expresses the orientation of l. The name *magnetic* arises because the angular momentum of a rotating charge (a current loop) is naturally associated with a magnetic moment.

s, spin quantum number. It was shown by Dirac in 1928, using a relativistic form of wave mechanics, that an electron must have an intrinsic spin angular momentum equal to $\sqrt{[s(s+1)]} \cdot h/2\pi$, where s always has the value $\frac{1}{2}$ for a single electron. Thus the spin is an intrinsic feature of the electron; it can never lose or gain spin angular momentum, although the direction or axis of the spin can change in a quantized manner.

The quantum numbers for the electrons of an atom or molecule can combine in a quantized manner and there is an extensive nomenclature to describe the various resultant quantum numbers. We need introduce only a few of the symbols. For historical reasons, the values of $l = 0, 1, 2, \ldots$, are described for atoms by the letters s, p, d, and for molecules by the equivalent Greek letters σ, π, δ. Thus a π-electron is one with an orbital angular momentum of unity.

In addition to these electron properties, a molecule, but not an atom, can have two other types of quantized energy denoted by quantum numbers.

Vibrational quantum number, v

There are obviously forces which keep the atoms of a molecule not far from their equilibrium positions. Wave mechanics shows that if these forces are simple harmonic (and this must be at least an approximation to the truth), the energy of vibration must be so quantized that

$$E = (v + \tfrac{1}{2})h\nu_0 \tag{2.5}$$

where ν_0 is the frequency of vibration of the parts of the molecule; note that ν_0 is not to be confused with the frequency of a photon, although we shall see (Chapter 6) that there is a simple relation between ν and ν_0.

Rotational quantum number, J

Equation (2.4) for the orbital angular momentum of an electron also applies to the angular momentum of a molecule rotating about an axis through its centre of gravity, but in this connection the quantum number is denoted by the letter J (sometimes also K) in place of l.

Molecular spectra

Energy differences between levels within each of the following groups, expressed in both cm^{-1} and electron-volts (see p. 8 and Table C at back of book), have the following orders of magnitude:

	cm^{-1}	electron-volts
electronic levels	$> 10{,}000$	$> 1 \cdot 2$
vibrational levels	200–3000	$0 \cdot 02$–$0 \cdot 35$
rotational levels	1–200	$0 \cdot 0001$–$0 \cdot 02$

The mean kinetic energy of a gas molecule is $3kT/2$, which at room temperature may be calculated, using the constants of Table A (back of book), to be about $0 \cdot 04$ eV. Comparing this with the last column above, we see that, at room temperature, excited rotational levels will occur; that virtually no excited electronic levels will exist, all the molecules being in the ground electronic state; and that there will be very little excitation into higher vibrational levels. See also the last column of Table C (back of book).

Subject to certain selection rules, combinations of quantum-number changes involving any combination of the above three types of energy level may occur. The region in which a spectrum occurs is determined by the quantum-number change involving the greatest energy difference. Thus molecular spectra are classified into three types, the contributions to the final wave-number from the electronic, vibrational, and rotational changes being respectively $\bar{\nu}_e, \bar{\nu}_v, \bar{\nu}_r$.

		Range in	
		Wave-number	*Wavelength*
		(cm^{-1})	(μm)
Electronic	$\bar{\nu} = \bar{\nu}_e \mp \bar{\nu}_v \mp \bar{\nu}_r$	$\gtrsim 10{,}000$	$\leqslant 1 \cdot 0$
Vibration-rotation	$\bar{\nu} = \bar{\nu}_v \mp \bar{\nu}_r$	10,000–200	1–50
Rotation	$\bar{\nu} = \bar{\nu}_r$	200–1	50–10,000

As one proceeds upwards in this table the spectra obviously become more complex, as is illustrated by Fig. 6.6 (vibration-rotation spectrum of HCl) and Fig. 3.5 (electronic spectrum of benzene).

Electronic absorption spectra

An electronic absorption band system for oxygen (O_2) is shown in Fig. 6.1, page 108. The whole of the system is concerned with one electronic change, accompanied by a jump of vibrational quantum number $v'' = 0$ to $v' = 8, 9, \ldots, 20$, as noted in the figure. The fine structure is due to the various changes of rotational quantum number. Fig. 3.5, page 32, shows a similar band system for benzene vapour; the rotational structure is more complicated because the molecule can in this case rotate about different axes having different moments of inertia.

In Chapter 6 the structures of electronic absorption bands will be considered further, together with vibration-rotation and pure rotation spectra.

Spectra of liquids and solids lose their fine structure because of the disturbing effects of neighbouring molecules; this is discussed further on page 32.

Fluorescence

After a molecule has been excited to a higher energy level by absorption of radiation, it may return to the ground state by collision processes, the absorbed energy finally appearing as an increase of temperature. But at least some of the excited molecules may return to lower energy levels by the emission of radiation which is named fluorescence radiation.

The mechanism involved in the emission of fluorescence radiation in the visible and ultra-violet is illustrated in Fig. 2.1; on the left is depicted the lowest electronic level E_1 and an excited energy level E_2. On the right are shown the sets of slightly higher levels due to the various vibrational levels with vibrational quantum numbers $0, 1, 2, \ldots,$ that can be associated with each electronic level. The processes involved in the emission of fluorescence radiation are as follows:

1. Excitation by absorption of radiation to one or other of the vibrational levels associated with E_2 (column A of the figure).

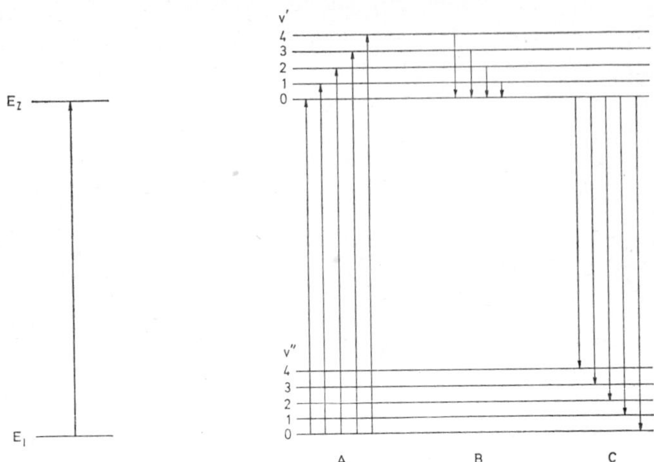

FIG. 2.1. Energy level diagram illustrating the processes involved in emission of fluorescence radiation. For detailed explanation, see text.

2. Collision degradation to the lowest vibrational level $v'=0$ of the electronic state E_2 (column B).

3. Return (column C) to one of the vibrational levels associated with the ground electronic state E_1, accompanied by the emission of fluorescence photons, which will obviously be of longer wavelength (lower wavenumber) than the original absorbed radiation. If the spacings of the vibrational levels associated with each electronic level are similar, it is fairly obvious that the spacings in the structure of the absorption band and the fluorescent band will be similar. In fact the two patterns will be mirror images, the reflection plane corresponding to the transition from $v'=0$ to $v''=0$, commonly named the 0, 0 transition. This mirror imaging is illustrated in Fig. 6.2 (page 110).

The Franck-Condon principle

Molecular vibrations arise from the restoring force—increased potential energy—set up when the parts of a molecule move from their equilibrium position, as at A in Fig. 2.2. The curves of this figure are unsymmetrical since there is a limit to the closeness of approach of two parts of a molecule; but the separation of two parts can increase indefinitely until the molecule becomes dissociated.   In two different electronic states the equilibrium distances might be equal as in Fig. 2.2(*a*), or unequal as in Fig. 2.2(*b*); this latter means that in the excited electronic state the molecule is more loosely bound together, the interatomic attraction is reduced.   The horizontal lines indicate the quantized vibrational energy levels; note that they are not equally spaced in energy, because, in contrast to the simple treatment leading to equation (2.5), the restoring forces, at least for the higher energy levels, are not exactly simple harmonic.

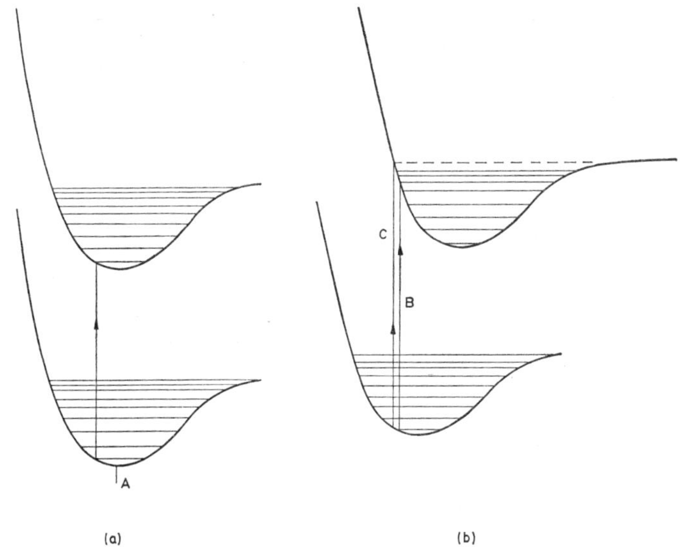

FIG. 2.2. Diagrams to illustrate the Franck-Condon principle. *Ordinates*: potential energy.   *Abscissae*: distance between the parts of a molecule.   Horizontal lines represent quantized vibrational energy levels.

The atoms spend relatively more time near the extremes of the vibration where the velocity is instantaneously zero, so that an electronic transition is more likely to occur at this phase of a vibration.   According to the Franck-Condon principle, the atoms have no time, during the short time of an electronic transition, to change their positions appreciably, so that the most probable transitions are vertical ones in Fig. 2.2.   Bearing in mind that at laboratory temperatures most molecules are in the lowest vibrational state, the most probable transitions are from $v'' = 0$, as shown for the excitations indicated by arrows in Figs. 2.2(a) and (b). The principle may be used in a much more quantitative way than suggested by this elementary discussion, and some applications are given in Chapter 6.

CHAPTER 3

The Laws of Absorption

QUANTITATIVE measurements of absorption are based on two fundamental laws relating the intensities of incident and transmitted radiation. The symbols and names used are summarized in Table B (back of book) and the relative status of the alternatives has already been discussed on page 6.

Lambert's Law

This was enunciated by Lambert some few years after its first statement by Bouguer in 1729 and is sometimes called the Lambert-Bouguer law. It states that the proportion of radiation absorbed by a substance is independent of the intensity of the incident radiation. The law can also be expressed in the form that each successive layer of thickness dl of the medium absorbs the same fraction dI/I of the radiation of intensity I incident upon it, i.e. $dI/I = -\mu\, dl$, where μ is a constant. This expression becomes on integration :

$$I = I_0\, e^{-\mu l} \qquad (3.1)$$

where I_0 is the intensity of radiation incident on a layer of thickness l and I is the emergent intensity. We may also write (3.1) as :

$$I = I_0 \cdot 10^{-Kl} \qquad (3.2)$$

or
$$\log_{10}(I_0/I) = Kl \qquad (3.3)$$

where
$$K = \log_{10} e \cdot \mu = 0.4343\, \mu \qquad (3.4)$$

Here μ is the absorption coefficient and K is the extinction coefficient or absorbance index.

Another important quantity A is defined by

$$A = \log_{10}(I_0/I) = \log_{10}(1/t) = Kl \qquad (3.5)$$

where $t = I/I_0$ is the fraction of radiation transmitted ; A is the quantity directly determined from a measurement, and instrument scales are often engraved with this unit. Older names for A are optical density and extinction, these terms being first introduced by Hurter and Driffield (1890) in connection with their researches on the blackening of photographic plates. The newer name for this quantity is absorbance.

Beer's Law

Beer's law (Beer, 1852) was based on quantitative and photometric observations on the absorption of red light by aqueous solutions of various inorganic salts. It states the relationship between the intensities of the incident and transmitted radiations in a different way, namely that the absorption depends only on the number of absorbing molecules through which the radiation passes. If the absorbing substance is dissolved in a non-absorbing medium, the optical density will then be proportional to the concentration of the solution—although Pfeiffer and Liebhafsky (1951), in a review of Beer's contribution, point out that he did not in fact state this simple corollary.

Beer's law can thus be combined with Lambert's law in the form :

$$A = \log_{10}(I_0/I) = acl \qquad (3.6)$$

or

$$I = I_0 \cdot 10^{-acl} \qquad (3.7)$$

where a is the extinction coefficient for unit concentration and is named the specific extinction coefficient and, in the newer systems, absorptivity or absorbancy index. Equations (3.6) and (3.7) are often referred to as the Beer-Lambert law, or more simply as Beer's law.

The numerical value of a obviously depends on the units of concentration and length. When the concentration is expressed in gram-molecules per litre and the length in centimetres, a special symbol ϵ is used instead of a, and the quantity is called the molar or molecular extinction coefficient, molar absorptivity or molar absorbancy index.

With several successive layers of absorbing material, or of a solution containing several absorbing components, it is obvious that the total transmission will be the product of the transmissions of the several components, and hence that the optical density will be the sum of the optical densities of the several components :

$$A = A_1 + A_2 + \ldots = a_1 c_1 l_1 + a_2 c_2 l_2 + \ldots \qquad (3.8)$$

Hence the great importance and convenience of measurements expressed as optical density or absorbance.

In cases where the molecular weight of a substance is not definitely known, it is not possible to write down the molecular extinction coefficient, and in such cases it is now common practice to write the unit of concentration as a superscript and the unit of length as a subscript. Thus the statement

$$E_{1\,\text{cm}}^{1\%}\ 3250\ \text{Å} = 20$$

means that for the substance in question, at a wavelength of 3250 Å, a solution of length 1 cm and concentration 1 per cent (usually meaning percentage of weight to volume, i.e. 1 g of substance per 100 cm³ of solution), $\log_{10}(I_0/I)$ has the value 20.

Another term, the absorption index k, is sometimes used in theoretical considerations (see, for example, page 43). It is defined by the equation

$$I = I_0 \cdot e^{-4\pi k l n/\lambda} \tag{3.9}$$

where λ is the wavelength of the light for which the absorption is to be specified and n is refractive index. It is possible to convert k into ϵ by the simple relationship

$$k = 0.183\ c\lambda\epsilon$$

These terms are summarized in Table B at the back of the book.

If the area of a molecule (or that part of it which is effective in absorbing a given radiation) is known, it is possible to calculate the fraction of radiation that should be absorbed on transmission through a layer of molecules. Such area (σ) is called the absorption cross-sectional area, or more briefly the *absorption cross-section*. Conversely, σ may be calculated if the absorption (i.e. the molecular extinction coefficient ϵ) is known. To express values of absorption of optical radiations in terms of σ is the exception rather than the rule, although in other branches of physics (such as the absorption of γ rays, neutrons, etc., by atomic nuclei) the use of σ values is almost universal.

Consider a concentration of c gram-molecules per litre. The number of molecules in a square centimetre of a layer of thickness dl is $cN_A dl/1000$, where N_A is Avogadro's number, and the effective area for absorption is $\sigma c N_A \, dl/1000$. Hence the fraction absorbed is

$$- dI/I = N_A \sigma c \, dl/1000$$

which integrates to

$$\log_{10}(I_0/I) = 0.43 N_A \sigma c l/1000$$

Hence the molecular extinction coefficient is given by

$$\epsilon = (1/cl) \log_{10}(I_0/I) = (0.43/1000) N_A \sigma = 2.6 \cdot 10^{20} \sigma \qquad (3.10)$$

A measured value of ϵ may often be enormously less than given by this expression. Reasons for this are as follows:

(1) In a liquid or gas, only a fraction of the molecules may be in correct orientation to absorb the incident radiation.

(2) An electron concerned in a transition may be localized in one part of a molecule as further discussed in Chapter 6.

(3) A transition may be 'forbidden' by quantum-mechanical selection rules. This usually means that the transition is not completely forbidden but is improbable. In such cases the right-hand side of equation (3.10) is to be multiplied by a factor f, known as the oscillator strength of the transition. This factor is further discussed in Chapter 6, page 130. To quote one example, Ditchburn, Jutsum and Marr (1953) find by measurements that the absorption cross-section of Na atoms to radiation of wavelength 2412 Å is $12 \cdot 10^{-20}$ cm^2, which is considerably less than the area of the whole atom (approximately 10^{-16} cm^2).

Validity of the laws

Beer's law was tested long ago by Vierordt (1873), who devised the first method of absorption spectrophotometry with any claim to accuracy, and it was repeatedly confirmed by later workers. Nevertheless, despite the lack of sensitivity of the semi-quantitative methods that were generally employed before the development of modern spectrophotometers, apparent exceptions were reported from time to time, so that Hulbert (1917) stated that 'this "law" of Beer has been shown to be the exception rather than the rule'. But Beer's law is now regarded as having general validity and apparent failures simply mean that it has been tried under conditions in which it should not be expected to apply, as discussed in the next section.

It is important to note, however, that Lambert's and Beer's laws apply only for homogeneous, that is monochromatic, radiation. If the radiation is not of a single wavelength, its quality as well as its quantity will change on passing through an absorbing medium and the initial statement of the two laws will therefore no longer apply. This gives rise to difficulties in the practical application of Beer's law to spectrophotometric method. These difficulties, and measures that may be taken to overcome them, will be discussed below.

FACTORS LIMITING THE OPERATION OF THE LAWS

In using spectrophotometric methods for analysis one should always ascertain by direct test whether Beer's law is operative in a particular case, but the following theoretical considerations will be of help when considering whether the law is likely to hold and when deciding how technique may be modified so that the law may be applied.

Factors which may make Beer's law inapplicable are :

1. Chemical changes.
2. Dichroism.
3. Finite width of waveband for observation.
4. Pressure broadening.
5. Scattering media.

These factors will now be discussed in turn, with special attention to scattering of radiation caused by non-uniformity of specimen and reflection from scattering materials. The relation between specular reflection and absorption, methods of determining the total intensity of an absorption band, and the advantage in certain instances of low-temperature measurements, will also be considered.

CHEMICAL CHANGES

Various changes of absorption on dilution may be attributed to chemical changes. When examining solutions of cobalt salts, for example, departures from Beer's law may be explained by the fact that these salts form complexes and the composition depends on the concentration. Groh and Papp (1930) attributed the variations that occur in the absorption of solutions of iodine when they are diluted to the presence of I_6 molecules.

A common cause of the apparent total failure of Beer's law is association in solution, which makes the proportions of the types of molecules present dependent on the concentration. This phenomenon is common with compounds containing hydroxyl groups; it is more fully treated on page 122, but meanwhile the reader may refer to Fig. 6.9 and note that, for solutions of benzyl alcohol in carbon tetrachloride, the whole shape of the absorption curve changes with concentration. In very dilute solutions there are only monomeric molecules of alcohol ; as the concentration increases, greater proportions of associated molecules $(C_6H_5CH_2OH)_n$ are present, giving rise to a new band at somewhat longer wavelength. As a result, the intensity of the absorption band due to the monomeric form actually decreases with increasing concentration of alcohol. Similar variations with tartaric acid are explained by the fact that the solutions contain two forms of the acid molecule in equilibrium.

DICHROISM

Specimens in which the absorption depends on the direction of polarization of the radiation, as discussed on page 124, will obviously not obey Beer's law when measured in unpolarized radiation. This effect is discussed by Glick, Engstrom and Malmstrom (1951) and by Commoner (1949), who has observed it in measurements of nucleic acids in cells.

For the same reason materials exhibiting circular dichroism (see page 127) will not obey Beer's law. But Woldbye and Bagger (1966) show that deviations will not normally amount to more than 1 per cent of a measured value of optical density. They confirmed their calculations by measurements on dextro-camphor.

FINITE WAVEBAND

It has been emphasized that Lambert's and Beer's laws should be applied only to monochromatic light of a single wavelength, but in practice a single wavelength is an unattainable ideal. If the light source has a line spectrum, a single line of this has a waveband that has a small but finite width on account of Doppler effect and pressure broadening ; in many cases such a light source can be considered truly monochromatic. But if the light source has a continuous spectrum and a monochromator is used to isolate one waveband, the wavelength range included in this depends on the widths of the slits, which must be at least wide enough to give adequate intensity for measurement. The band width also depends on the optical perfection of the instrument—in particular on the resolving power and freedom from stray radiation. These last two effects are considered more fully on pages 62 and 67 respectively. Meanwhile we shall assume that instead of a single wavelength we have a finite waveband.

Suppose we have an absorbing substance in two concentrations of ratio, say, $4:1$, and the absorption curves for the two concentrations are as shown in Fig. 3.1a. If this substance obeys Beer's law, the ordinates of the two curves are everywhere in the ratio $4:1$. Now suppose the spectrophotometer isolates a waveband ab ; for each concentration it will give an effective optical density somewhere between the extreme values at wavelengths a and b. But in general the measured optical density will not be the mean of the extremes. If I_λ is the intensity of the incident beam, A_λ the absorbance, and S_λ the relative sensitivity of the receiver (eye, photocell, photographic plate, etc.), all at wavelength λ, then the transmitted intensity at this wavelength is $I_\lambda . 10^{-A_\lambda}$, and the response in the receiver is given by

$$\int_{\lambda_a}^{\lambda_b} I_\lambda S_\lambda . 10^{-A_\lambda} d\lambda$$

and the measured transmission is given by

$$\bar{t} = \frac{\int_{\lambda_a}^{\lambda_b} I_\lambda S_\lambda . 10^{-A_\lambda} d\lambda}{\int_{\lambda_a}^{\lambda_b} I_\lambda S_\lambda \, d\lambda} \tag{3.11}$$

and the measured absorbance is given by

$$\bar{A} = \log_{10} (1/\bar{t})$$

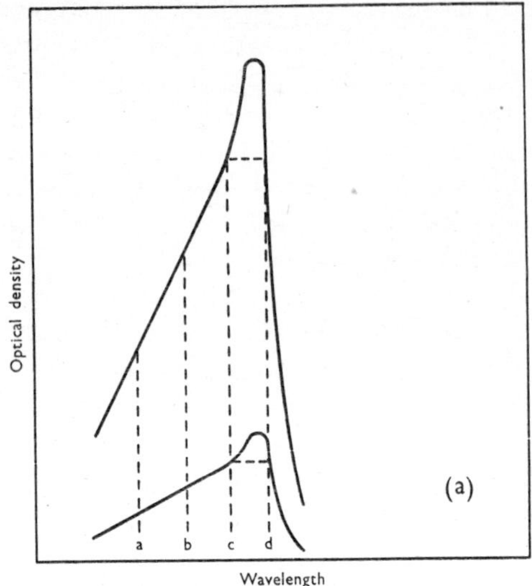

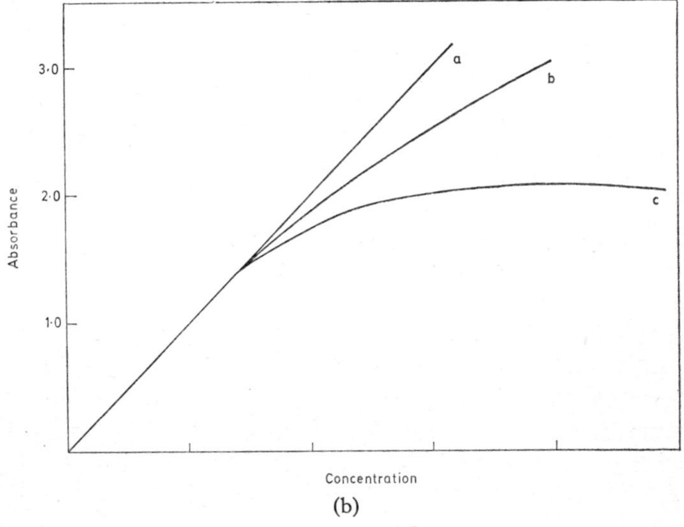

Fig. 3.1. Diagram illustrating the applicability of Beer's law. (a) Curves for two concentrations obeying the law: (b) curve a: Beer's law; curve b: apparent failure of the law, arising from finite slit widths—see text; curve c: the effect of 1% of unabsorbable stray radiation in the beam from a monochromator. (Fig. 3. 1b is reproduced from Lothian, 1963.)

With a finite waveband, the measured absorbance will therefore depend not only on the shape of the absorption curve but also on the intensity distribution in the light source and on the wavelength variation of the response in the receiver. (In the case of a measurement at what is effectively a single wavelength the value is independent of the means of measurement.)

Integrating in the above equation will enable us to calculate the effective optical density under specified conditions. If the integration is carried out (either mathematically when the variations of absorbance, incident intensity and receiver sensitivity with wavelength are simple mathematical functions, or empirically in other cases) it is found that the effective absorbance is less than the absorbance at the mean wavelength, both when the absorbance–wavelength curve is straight, as at *ab* in Fig. 3.1*a*, and when it is convex upwards as at *cd*. Between wavelengths *a* and *b* the smaller optical densities give greater intensities of transmitted light and therefore contribute more than the higher ones towards the mean; it is even more obvious that in passing over a peak, as between *c* and *d*, the mean absorbance for the waveband will lie below the peak value—and also that for a sharper maximum such as occurs at higher concentrations the difference between effective mean and peak absorbance will be increased.

It will be seen, if equation (3.11) is applied to a slope such as *ab*, that the difference between effective absorbance and absorbance of mean wavelength is increased when the absorbance range within the band *ab* is increased. The magnitude of this effect may be readily determined if we simplify calculations by three reasonable assumptions—that the receiver sensitivity is uniform over the waveband, that the incident intensity as transmitted by the monochromator is uniform, and that the absorbance varies linearly with the wavelength. The resulting table may be of some use in estimating whether Beer's law should be expected to hold under particular experimental conditions.

Absorbance range between extreme wavelengths	0·1	0·2	0·4	0·6	0·8	1·0	2·0
Difference between effective absorbance and absorbance of mean wavelength	0·001	0·005	0·015	0·034	0·060	0·092	0·332

When the concentration is increased, the absorbance range included in the measurement will be increased and, as the above table shows, the measured value will become increasingly less than that of the mean wavelength, and the graph of absorbance against concentration will be curved as shown in Fig. 3.1*b*. The effect of these changes becomes very important when narrow absorption bands have to be measured, and may be apparent when dealing with infra-red rotation-vibration bands. An example is described on pages 86 ff., together with the method of dealing with it.

In abridged spectrophotometry (see page 195), a relatively wide waveband may be isolated by means of a filter, and tests will show whether or not an absorbance so measured is proportional to concentration. But, even with proportionality, the measurements obtained must be regarded as quite empirical and applicable only to the set-up used. Measurement depends not only on the absorption curve of the substance being examined and the wave-length-transmission curve of the filter but also on the wavelength-intensity distribution of the light source and the sensitivity of the receiver.

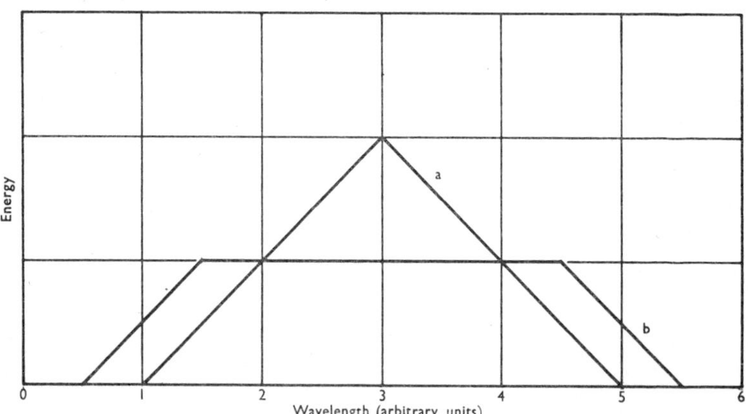

FIG. 3.2. Monochromator with finite slits. Energy-wavelength distribution of the radiation transmitted by a monochromator illuminated with an equi-energy continuous spectrum source. (*a*) Equal slits of 2 wavelength units each. (*b*) Unequal slits of 4 and 1 units. The energies transmitted are the same in each instance but in (*a*) 75 per cent of the transmitted radiation is contained in a 2-unit range and in (*b*) only 50 per cent.

Correcting measurements to zero slit width

Brodersen (1954) has given a treatment of slit-width effects in which he assumes that an absorption band is of Gaussian shape, that is,

$$A = A_{\max} \exp \{ -(\bar{\nu} - \bar{\nu}_0)^2 / 2\sigma^2 \} \qquad (3.12)$$

where σ defines the width of the band (the semi-width is $1\cdot17\sigma$ at half the maximum height). He also assumes that the monochromator is used with equal slits, in which case the energy transmitted by it, when plotted against ν or λ, gives an isosceles triangle (Fig. 3.2) whose semi-base $s = \varDelta\bar{\nu}$ is a measure of the slit width in wavenumbers. One of the things determined by Brodersen was

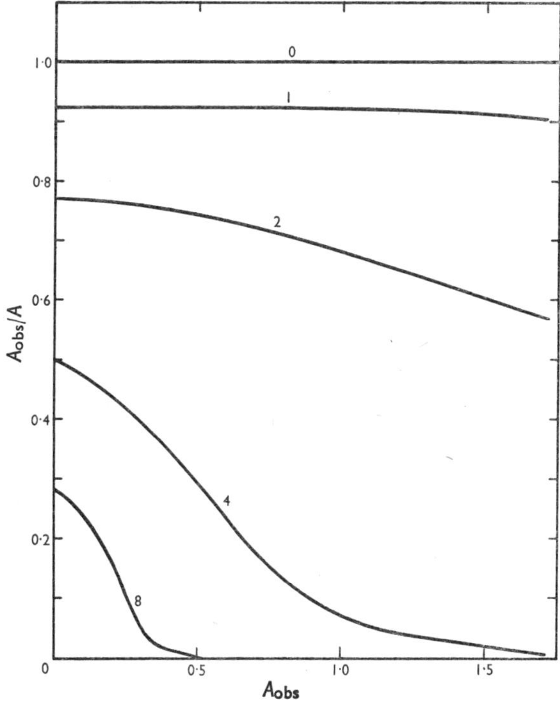

FIG. 3.3. Ratio of observed to true absorbance (ordinates) at the peak of an absorption band for various values of absorbance (abscissae), and for various values of the ratio of slit width to width of absorption band (s/σ) as marked on each curve. (From Brodersen, 1954.)

how the ratio of observed to true absorbance at the peak of the band depends on the absorbance and on the ratio s/σ; this is shown in Fig. 3.3, which is reproduced from his paper. A horizontal line on this diagram indicates that Beer's law holds; it is interesting to note that for $s/\sigma = 1$ the law holds fairly exactly, and yet the measured absorbances are only about 92% of the true values. It should be noted, therefore, that operation of Beer's law does not mean that true absorbances are being measured.

A curve measured with finite slits may be corrected to approximate to the true curve in a manner first described by Paschen (1897) and explained very clearly in a paper by Slater (1925). If the measurements are plotted as transmittance t against $\bar{\nu}$ or λ (Fig. 3.4) and if it may be assumed that the element of the curve within the waveband $2s$ passed by the monochromator is parabolic, then the correction at λ in first approximation can be made as follows: ordinates are drawn at $\lambda + s$ and $\lambda - s$ and a chord drawn where these cut the curve; the correction to the ordinate at λ is then equal to one-third of the ordinate difference between chord and curve at λ; i.e. $\frac{1}{3}$ of pq or $p'q'$. Paschen describes higher-

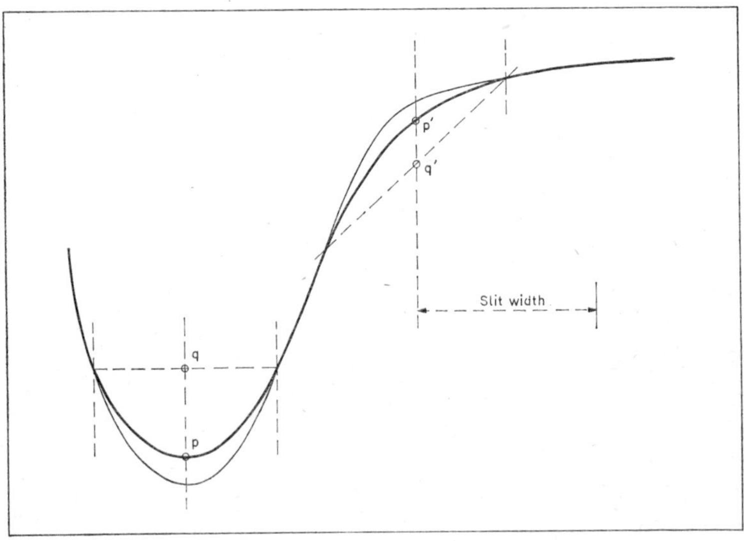

FIG. 3.4. Correction for finite slit width. —observed;—corrected. For details see text.

order corrections, but the method ignores the variation of energy from the monochromator within the range $2s$ and it is doubtful whether such further correction is justified.

An alternative and useful treatment is that of Willis (1951), who assumes that an absorption band has the Gaussian form

$$(1-t)=(1-t)_{\max} \exp \{-(\bar{\nu}-\bar{\nu}_0)^2/a^2\}$$

and that the energy passed by the monochromator may be expressed by a similar function

$$I\bar{\nu}_1=I\bar{\nu} \exp \{-(\bar{\nu}-\bar{\nu}_1)^2/b^2\}$$

which is not very different from the triangular form assumed above. It can then be shown that at the peak of the band

$$(1-t)_{\max \text{ obs}}=(1-t)_{\max \text{ true}} \cdot a/(a^2+b^2)^{\frac{1}{2}}$$

and the true width a of the band is related to the observed width a' by the relation

$$a^2=(a'^2-b^2) \tag{3.13}$$

Willis has confirmed these conclusions with several sets of measurements in the infra-red.

It must be emphasized that no treatment of observations can reveal detailed structure within the waveband $2s$ used to obtain the measurements. A useful bibliography of papers on slit width effects is given by Brodersen (1954).

PRESSURE BROADENING

When a gas at low pressure is absorbing, the bands are sharp and narrow. As the pressure is increased, so that the molecules are closer together and collisions are more frequent, the bands become broader; when the substance is liquid, the molecules being extremely close together and the collisions much more frequent than in any gas, the individual bands may be so broadened that they are completely lost and only the envelope of the band obtained. This effect may be seen by comparing the familiar absorption spectrum of benzene solution near 2600 Å (Fig. 9.16, page 199) with that of benzene vapour (Fig 3.5), which shows much more structure. The absorption bands of vapour in Plate 1 opposite page 32 and Fig. 9.1 on page 177 may also be contrasted with the bands of liquid cresol in Fig. 5.1 on page 83.

Classical theory could explain this broadening by saying that when a molecule suffers a collision while it is absorbing or emitting a train of waves the process of absorption or emission is broken off. The wave-train (absorbed or emitted) of finite length is equivalent to a series of waves of different wavelengths, i.e. it becomes a wave of finite spectral width.

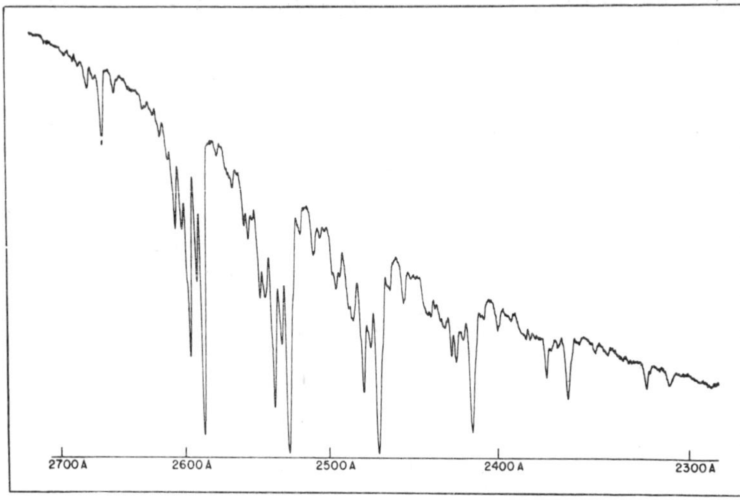

FIG. 3.5. Absorption of benzene vapour in the ultra-violet, recorded on a Perkin-Elmer single-beam spectrophotometer (Model 112) with hydrogen lamp and photomultiplier. Crystal quartz prism, 0·050-mm slits, and recording time of 54 minutes.

From the quantum mechanical point of view, the electrons or atoms are in a field of force determined not only by their own molecule but modified in an irregular manner by the neighbouring molecules. Thus the energy levels become indefinite and, averaged over a large number of molecules, become broadened into bands. This can occur when a molecule meets another whose rotational energy differs by one quantum (see Chapter 6 for the quantum selection rules). This sort of collision is most common for molecules with rotational energy near the equipartition value kT, and Foley (1946) has shown that, for the vibration-rotation spectrum of HCN, the greatest pressure broadening occurs for the

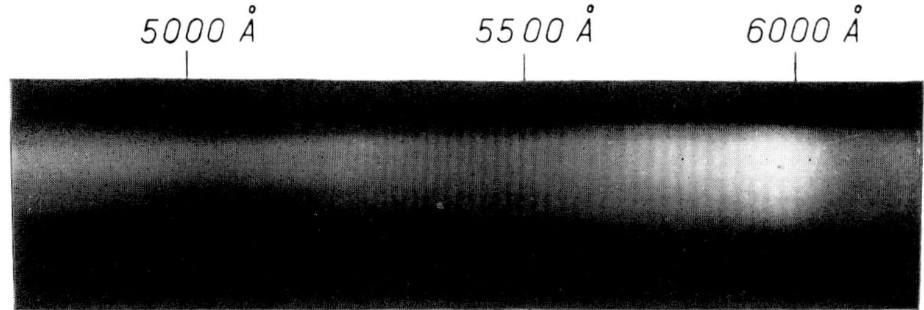

PLATE IA. Electronic absorption spectrum of iodine vapour. At $\lambda \leqslant 4995$ Å, the band structure disappears and the absorption becomes continuous (see p. 107).

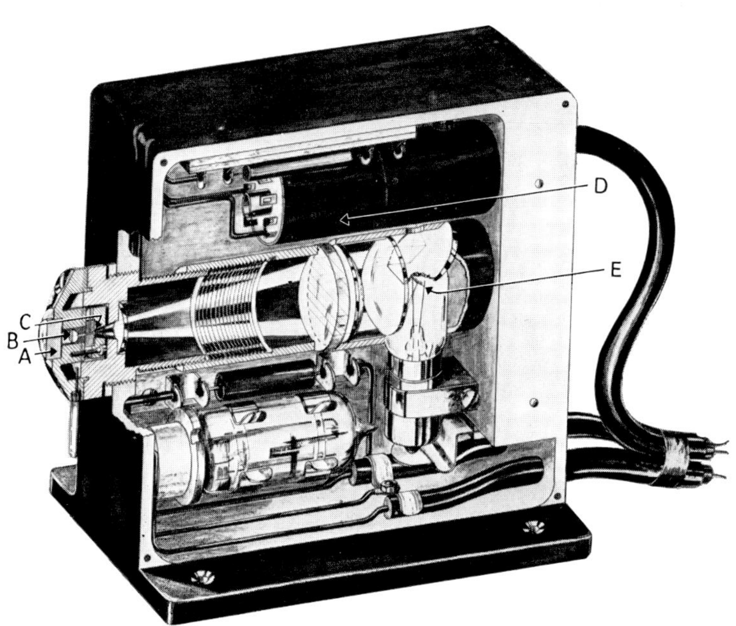

PLATE IB. Golay infra-red detector. For description, see p. 170.

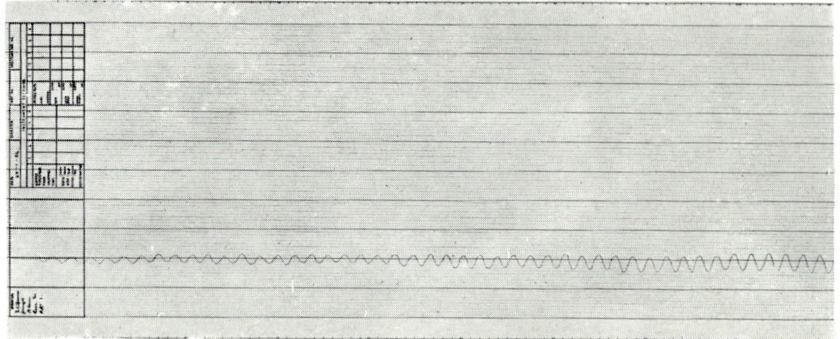

PLATE 2A. Wavelength calibration by interference fringes. A record obtained with a Unicam infra-red spectrophotometer using two KBr plates with an air space of about $\frac{1}{4}$ mm between them. Wavelength range 5–15 μ. Slit adjusted to constant energy. Recording time of 6 minutes.

(a)

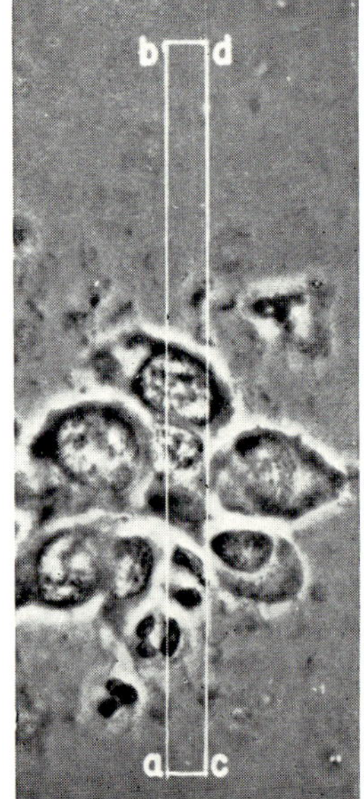

PLATE 2B. (a) Phase-contrast photomicrogram ($\times 600$) of cervical smears fixed in formalin-saline. (b) Microspectrogram ($\times 600$ vertically and $\times 200$ horizontally), obtained with a line source of radiation, of the area marked in (a). (Mellors, 1950.)

(b)

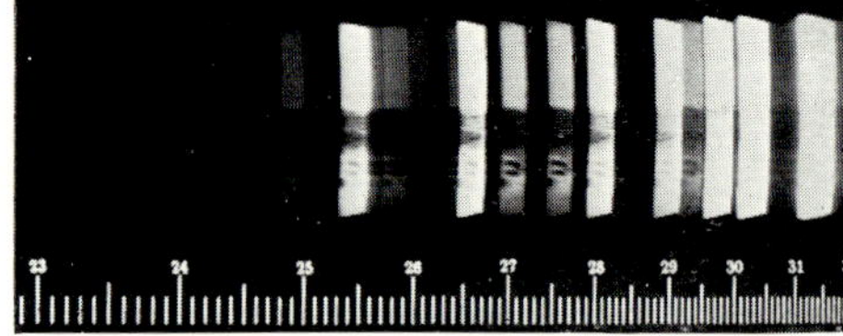

quantum number $j=7$, being less than a quarter of this maximum value for $j=0$ and $j=22$.

This pressure broadening causes an important departure from Beer's law, the broadening being accompanied by an effective increase in intensity of absorption. The effect was observed as early as 1889 by Ångstrom, and between 1909 and 1913 a large number of measurements were made by Eva von Bahr, who studied the principal absorption maxima of a number of gases in the region transmitted by rocksalt. An example in Fig. 3.6 shows how absorbance of methane at $\lambda=7\cdot65\,\mu$, at a constant partial pressure of $13\cdot15$ cm Hg, increases by about 50 per cent when a non-absorbing gas such as oxygen is added to increase the total pressure; the magnitude of the effect depends somewhat on the nature of the added gas.

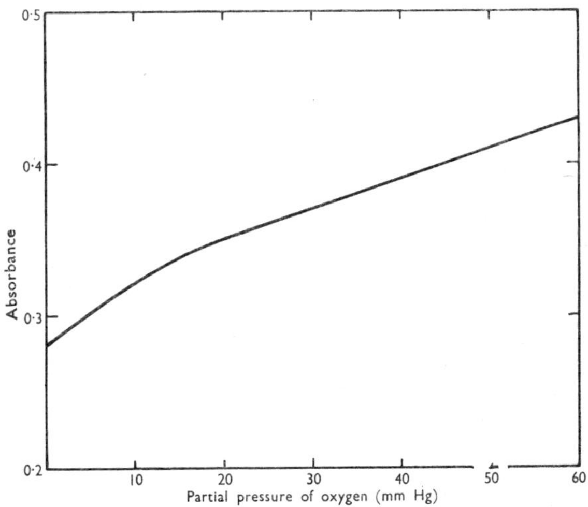

FIG. 3.6. Absorbance of methane at $\lambda=7\cdot65\,\mu$m. Constant partial pressure of methane in presence of varying partial pressures of oxygen (0–60 mm Hg).

Here the whole basis of Beer's law and its use in analysis is lost ; not only is the absorbance of methane not proportional to the number of methane molecules through which the radiation passes, but it depends on the nature and quantity of other non-absorbing gases present. It is found, for instance, that for a mixture of

c

methane with oxygen, at a constant total pressure of 80 cm Hg, the absorbance (at $7\cdot65\,\mu$ and cell length of $9\cdot5$ cm) is not proportional to the partial pressure of methane but increases more slowly at higher pressures :

Partial pressure of CH_4 (cm Hg)	$2\cdot5$	$5\cdot0$	$10\cdot0$	$20\cdot0$
Optical density	$0\cdot2$	$0\cdot25$	$0\cdot35$	$0\cdot45$

The above examples are taken from a paper by Coggeshall and Saier (1946), which also shows how practical analysis can be carried out in certain cases in spite of the breakdown in Beer's law; e.g. analysis of flue gases CO_2, CO and SO_2, making use of their infra-red absorption bands at $14\cdot8\,\mu$, $4\cdot65\,\mu$ and $7\cdot35\,\mu$ respectively.

In addition to this broadening of the energy levels, and hence of the absorption bands that arise from transitions between them, the energy levels may suffer some systematic disturbance causing a shift as well as a broadening of the band when the pressure is increased. Thus Oksengorn (1955 and 1956) has found that, when the vapours of benzene, toluene, or xylene are mixed with nitrogen at pressures up to 1000 atmospheres, the bands near 2600 Å are shifted about 6 Å towards longer wavelengths as well as being broadened. Foley (1946) finds a shift that is just measurable in the vibration-rotation bands of HCN when the pressure is increased. A very useful summary of the effects of both finite waveband and pressure broadening is to be found in a paper by Nielsen, Thornton and Dale (1944).

SCATTERING MEDIA

Radiation will be diffused or scattered by a medium which has a physical structure not very small compared with the wavelength of the radiation. This occurs, for example, with colloidal solutions, a precipitate in suspension, a cloud of water drops, an emulsion, a sheet of paper, etc. The transmission of such a medium will depend not only on its absorbing properties, but also on its scattering properties, which are determined by the refractive index of the particles composing it as well as by their size relative to the wavelength of the radiation ; the measured transmittance will depend also on the amount of the scattered radiation that is picked up by the receiver, i.e. on the design of the optical system used. The subject is a large one and has only been fully investigated for some special cases ; aspects which are relevant to this book will now be briefly discussed.

Effect on applicability of Beer's law

When a nearly parallel beam from an approximately point source at the principal focus of a lens is used and, after being transmitted through the specimen, is focused on a detector that is also approximately a point, most of the scattered light will be scattered out of the measured beam. Thus the measured transmittance will depend not only on the absorption coefficient (or molecular extinction coefficient) but also on a scattering coefficient, and the sum of these coefficients will occur in equations such as (3.6), etc. Beer's law may therefore be expected to hold for small amounts of scattering.

If, on the other hand, the detector picks up most of the radiation scattered in directions near that of the incident beam, the transmittance will depend mostly on the absorption coefficient ; the process of multiple scattering will tend to increase the transmittance of larger concentrations, and so Beer's law will become inoperative at much smaller concentrations than in the previous circumstances. Such effects were found by Rose and French (1948) in measurements on suspensions of powders.

When the concentration of scattering material is greatly increased, the transmittance will tend to fall again below the Beer's law value ; the longer path of the radiation through the medium caused by the multiple scattering will give rise to increased absorption. Such an effect has been found and explained for photographic emulsions, the absorbance being up to fifteen times greater than for a homogeneous layer of silver bromide—see Pitts (1954) and Farnell (1954). This phenomenon is useful in enhancing weak bands as described on page 55.

Absorption measurements on scattering materials

Since the transmittance in a parallel beam depends on both absorption and scatter, the wavelength variations of the latter may well mask the normal absorption curve. For example, curve 1

of Fig. 3.7 is an absorption curve obtained from measurements on a
suspension of blood corpuscles in saline, using a parallel beam of
radiation.   The γ band of haemoglobin near $\lambda = 4200$ Å appears
reversed, i.e. the normal absorption maximum appears as an
absorption minimum ; this happens because, as more fully

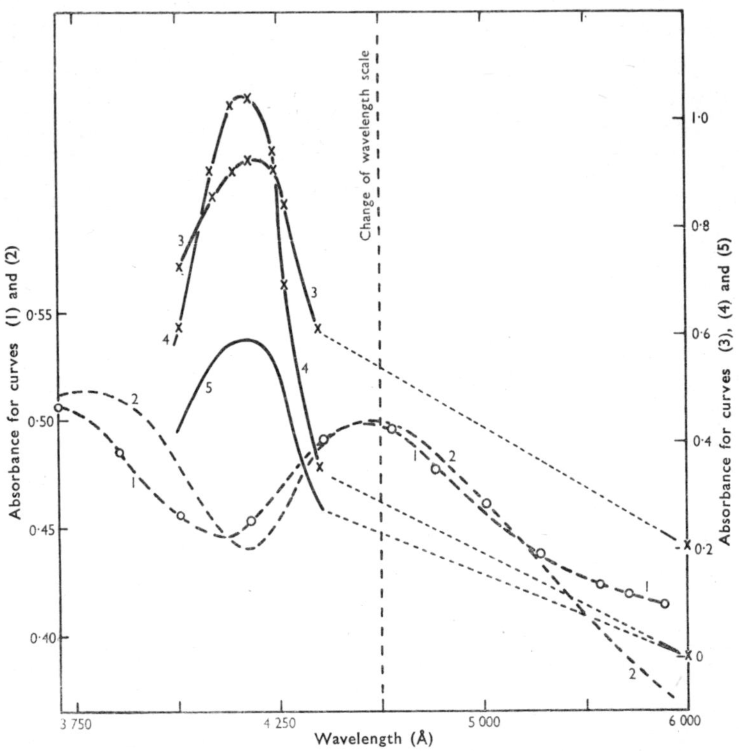

FIG. 3.7. Absorption of blood. Measured (curve 1) and calculated (curve 2)
absorption of a suspension of blood corpuscles in normal saline when using
a parallel beam of radiation. Measured (curve 3) and calculated (curve 5,
from an equation similar to 3.14) absorption of a suspension of corpuscles
when using a wide-angle beam ; for a discussion of the discrepancy between
curves 3 and 5 see Lothian and Lewis (1956), from which this figure is taken.
Curve 4 shows absorption after haemolysis of specimen used for curve 3.

explained by Lothian and Lewis (1956), the destructive interference between the waves passing through the particle and beside
it does not occur when the radiation incident on the particle is

absorbed. The result with a beam not made sufficiently parallel to show this reversal will lie between that just described and the normal absorption maximum ; in other words the γ band may well be effectively absent with corpuscular blood, as noted by early workers such as Adams, Bradley and McCallum (1938). Keilin and Hartree (1941) suspected that this non-appearance was an optical rather than a chemical effect and Robinson (1941) found the normal band when using a detecting arrangement that picked up the forward scattered radiation. One useful way of collecting a considerable proportion of the scattered radiation in order to investigate the absorption spectrum of diffusing materials is to focus an image of the specimen on the slit of the mono-chromator as is done in some commercial instruments (for an

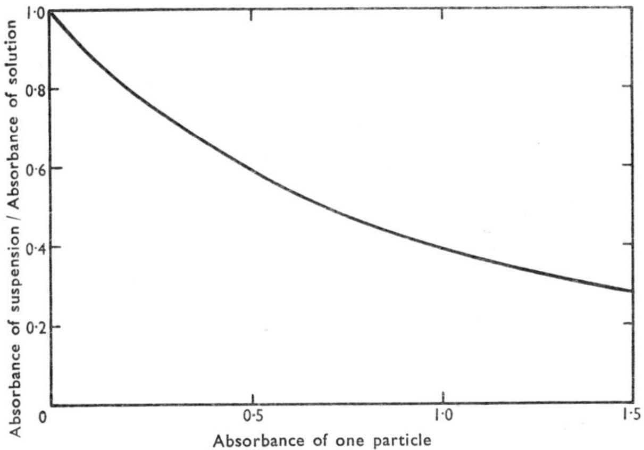

FIG. 3.8. Absorbance of a suspension of particles relative to that of a solution of the particles as a function of the absorbance of a single particle. Plotted from equations (3.14) and (3.15).

example see Fig. 9.4 on page 180). Another way is to use a diffusing surface immediately following the specimen—such as a ground exit window on the specimen cell. Such a method is used in determining the effective density for contact printing of photographic negatives. Barer (1955) has developed a technique for measurements on absorbing particles in which the scattered radiation is reduced by immersion in a medium of similar refractive index.

Strongly absorbing particles

If the particles of a diffusing medium are so large or so strongly absorbing that passage of the radiation through a single particle causes appreciable absorption, then, even when an arrangement that collects most of the scattered radiation is used, the absorption curve will differ in shape from that of the homogeneous material ; the peaks will be depressed.

To understand the nature and magnitude of this effect, consider a beam of radiation, of intensity I per unit area, passing through a length dl of a suspension containing n particles per unit volume, each of thickness t in the direction of the beam and of area S perpendicular to the beam. The intensity $I - dI$ reaching the other side of the layer is equal to the sum of the radiation passing through the clear spaces between the particles and the radiation transmitted through the absorbing particles of extinction co-efficient K ; i.e.

$$I - dI = I(1 - nS \,.\, dl) + InS \,.\, dl \,.\, 10^{-Kt}$$

After rearrangement, and integration over a definite layer of thickness l, this expression becomes

$$A_{\text{suspension}} = \log_{10}(I_0/I) = \log_{10} e \,.\, nSl(1 - 10^{-Kt}) \qquad (3.14)$$

If the particles are now dispersed uniformly in the same total volume, the absorbing material is diluted $1/nSt$ times, so that the extinction coefficient is now $nStK$ and the absorbance is

$$A_{\text{solution}} = nStKl \qquad (3.15)$$

Examination of these two expressions shows that they approach equality when Kt is small and that, as Kt increases, the absorbance of the suspension becomes increasingly less than that of the solution (see Fig. 3.8). Similar treatments of this effect are given by Stewart (1955) and Jones (1952).

This explains why the γ band for haemoglobin in unhaemolysed blood is considerably flatter than that for haemolysed blood (Fig. 3.7, curve 3); the absorbance of a single blood corpuscle in its smallest dimension is about 0·5 to 0·7 at the peak of this band. This effect is also of importance in measuring the absorption of powders by the KBr pellet or mull techniques (see page 171);

the more strongly absorbing the material the smaller it should be ground to obtain easily interpreted measurements.

Magnitude of scattered radiation

As a first approximation, one may suppose that all the radiation incident on a particle is refracted away from its original direction ; when measurements are then made with a parallel beam as discussed on page 150, even transparent particles will behave as if they are perfectly opaque, and the absorbance may be deduced from equation (3.14) by putting $K=\infty$, i.e.

$$A=\log_{10} e \cdot nSl \qquad (3.16)$$

This expression is often used to determine nS, which is proportional to the surface area of the particles ; if n or S is known and if nSl, which is proportional to the total mass, is found by weighing, then both S and n may be determined. Examples of such measurements are given by van Someren and Rollason (1945), Skinner and Boas-Traube (1947) and Rose and French (1948).

Such measurements must be accepted, however, with great reserve. Not only is the radiation travelling in the forward direction modified by diffraction effects at the particle boundary, but the radiation transmitted through the particle emerges with a phase retardation and suffers interference from the radiation passing near the particle. These effects may be expressed by means of a multiplying factor on the right-hand side of equation (3.16) known as the scattering-area coefficient or the total scattering coefficient, which may have a value up to 4. Approximate calculations of the numerical values of this factor may be made, using the optical principles just mentioned—see, for example, Van de Hulst (1957) and also Chappel and Lothian (1951). But the principles of physical optics are not strictly applicable when the dimensions of an obstacle are comparable with the wavelength ; an exact solution for spherical particles using electromagnetic theory has been given by Mie (1908) and numerical values computed by Lowan (1949), Gumprecht and Sliepcevich (1951), Irvine (1965) and others. Computations for cylindrical particles have been made by Van de Hulst (1950).

Values of the scattering-area coefficient depend on the ratio of particle radius r to the wavelength of the radiation and on the refractive index and the absorption of the particles. In general, graphs of scattering-area coefficient plotted against r/λ show a series of maxima and minima, and approach asymptotically the value 2 for large values of r/λ. The calculations assume that the measuring beam is ideally parallel ; experimental measurements approximating to these conditions agree reasonably well with the calculations—see, for example, Andreasen *et al.* (1950), Paul and Jones (1951) and Lewis and Lothian (1954). Goulden has used these principles to determine the sizes of fat globules in milk (see Chapter 7, page 149).

REFLECTION FROM SCATTERING MATERIAL

The radiation reflected from or transmitted through intensely scattering material is determined by a complicated process of multiple scattering and absorption. A very general treatment is that of Kubelka and Munk (1931) and Kubelka (1948), who deduce a number of formulae for a layer whose inhomogeneities are small compared with the total thickness of the layer.

For a layer so thick that the reflection does not depend on the nature of the backing material, this theory leads to the relation

$$2R/(1-R)^2 = \sigma/\mu \tag{3.17}$$

Here R equals the reflection, μ is the absorption coefficient already defined on page 19, and σ is a scattering coefficient defined in an analogous manner. For a mixture of materials 1, 2, etc. in concentrations c_1, c_2, etc., equation (3.17) becomes

$$2R/(1-R)^2 = \frac{c_1\sigma_1 + c_2\sigma_2 + \cdots}{c_1\mu_1 + c_2\mu_2 + \cdots} \tag{3.18}$$

Colour of mixed pigments

Equation (3.18) has been used by Duncan (1940 and 1965), who in the later paper reviews its use for the development of methods of colour prediction, and quotes some typical results. Tables of Kubelka-Munk functions to facilitate calculations are given by Judd and Wyszecki (1963), who devote some 20 pages to discussing the application of the Kubelka-Munk function to predicting the colours of mixtures. Modified tables suitable for use with paints, where reflection at the air-medium interface must be allowed for, are given by Duncan (1962). A 'computer' to facilitate the calculations has been described by Davidson and Hemmendinger (1958) and by Davidson (1962); reflection measurements at sixteen wavelengths are fed into the instrument, which has stored in it the μ and σ coefficients for a number of pigments. The concentrations of these are adjusted by rotation of knobs until a spectrophotometric match at the sixteen wavelengths is indicated by a series of dots all at the zero line on the screen of a cathode-ray oscillograph.

A simpler function than that of equation (3.17) has been suggested for use with certain types of dyed fabrics by Atherton (1955), viz.

$$1/R - 1/R_0 = \mu_1 c_1 + \mu_2 c_2 + \ldots \tag{3.19}$$

where R_0 is the reflection of the undyed fabric. Another more complicated function

$$F(R) = (1 - R^2)/(R - R_0)(1 + kR)$$
$$= F(R_0) + \mu_1 c_1 + \mu_2 c_2 + \ldots \tag{3.20}$$

has been found by Alderson, Atherton and Derbyshire (1961) to be of more general use than that of expression (3.19).

These last authors show how by storing, in a digital computer, wavelength-by-wavelength values of expression (3.20) for a number of dyes, dye compositions to match a given colour may be predicted. The colour of a specimen required to be matched is measured in trichromatic units by a colorimeter [several such instruments are described by Wright (1964)] and the colour specification fed into the computer. This method has been developed by ICI Dyestuffs into a system known as instrumental match prediction; it is described by Dawson, Derbyshire and Lemin (1965).

Chromatographic estimations

Equation (3.18) may also be used to determine the concentration of an adsorbate such as may be obtained in chromatographic analysis. If it may be assumed that the adsorbing material (o) is white ($\mu_0=0$) and that the adsorbate (1), being molecular, is non-diffusing ($\sigma_1=0$), equation (3.18) becomes

$$(1-R)^2/2R = c_1\mu_1/c_0\sigma_0 \tag{3.21}$$

The function $(1-R)^2/2R$ is thus proportional to the concentration of the adsorbate and represents a property similar to that of optical density in solutions. Schwuttke (1953) confirmed the proportionality except for wavelengths at which the adsorbate has only small absorption.

RELATION BETWEEN SPECULAR REFLECTION AND ABSORPTION

The theory of specular reflection from a polished surface is set out in text-books of physical optics and of electromagnetic theory. The fundamental equations were first developed by Fresnel for transverse waves travelling in an elastic medium and are still known as the Fresnel equations. Those relating to reflection, in which we are interested, are:

$$R_\perp = \sin^2(\theta-\theta')/\sin^2(\theta+\theta') \tag{3.22}$$

$$R_\parallel = \tan^2(\theta-\theta')/\tan^2(\theta+\theta') \tag{3.23}$$

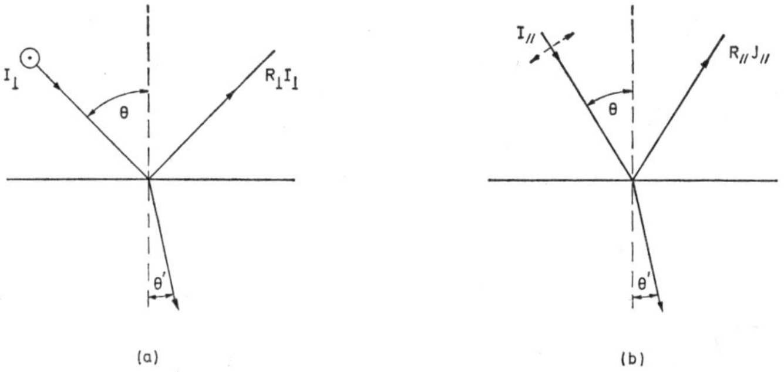

FIG. 3.9. Reflection of plane polarized radiation. Electric vector of wave (a) perpendicular and (b) parallel to plane of incidence.

where θ, θ' are the angles of incidence and refraction (Fig. 3.9) and $R_\perp$, $R_\parallel$ are reflection factors for radiation plane polarized with the electric field perpendicular and parallel respectively to the plane of incidence. Sometimes $R_\perp$, $R_\parallel$ are replaced by

R_s, R_p respectively (from the German *senkrecht* = perpendicular and *parallel* = parallel). These equations may be combined with the law of refraction:

$$n \sin \theta = n' \sin \theta' \qquad (3.24)$$

where n, n' are the refractive indices on each side of a boundary. For transparent media, these equations lead easily (1) to the well-known expression for normal incidence:

$$R_\perp = R_\| = (\mu - 1)^2/(\mu + 1)^2 \qquad (3.25)$$

where $\mu = n/n'$, the relative refractive index; and (2) to the existence of total internal reflection.

For absorbing media in which we are interested here, the refractive index may be written as a complex number, with real and imaginary parts

$$n(1 - ik) \qquad (3.26)$$

where k is the absorption index defined in Table B, and $i = \sqrt{-1}$. If this expression is combined with the equations (3.22–4) they may be developed into expressions for the reflection co-efficients of absorbing media and for the phase difference between the parallel and perpendicular polarizations of the reflections. The mathematics is involved and the reader is referred for details to the sources quoted below for the individual methods.

Measurement of some properties of the reflected radiation forms the most useful method of measuring both n and k, known as the optical constants, for strongly absorbing materials that it is not possible or desirable to dilute as a solution, e.g. metals, metalloids, etc. By substituting the results of a pair of measurements into two simultaneous equations, n and k may be determined. Some alternative methods are as follows.

1. When the incident radiation is plane polarized the reflected beam is in general elliptically polarized, and a determination of the parameters of the ellipse (the ratio of the axes and their orientation to the plane of reflection) at one angle of reflection is sufficient to permit calculation of n and k. A photographic method has been developed by Bor (1952) in which all the information required to determine n and k for a range of wave-lengths is obtained on one plate. Archard, Clegg and Taylor

(1952) have described a method for determining the ellipse parameters photoelectrically. Conn and Eaton (1954a) and Beattie and Conn (1955) have made measurements for wavelengths in the infra-red up to 12 μm, using selenium polarizers before and after the reflecting specimen. The relevant equations deduced from expressions (3.22), (3.23), (3.24) and (3.26) are involved and their numerical solution is long. These workers show that graphical solution is simpler. Conn, Clark and Parry (1968) describe an apparatus which gives ratios of reflection intensities for different polarizations in a form particularly convenient for graphical solution.

2. The method of Simon (1951) was to measure the reflected intensity at two angles of incidence.

3. D.G. Avery (1952) has described a method in which the ratio of the reflected intensities, for incident radiation plane polarized in and perpendicular to the plane of reflection, is determined for two different angles of incidence θ. The equations relating n, k and $R_\perp/R_\parallel$, which may be solved graphically, are

$$n^2(1 - ik)^2 - \sin^2 \theta = (a - ib)^2$$

$$R_\perp/_\parallel R = \frac{a^2 + b^2 - 2a \sin \theta \tan \theta + \sin^2 \theta \tan^2 \theta}{a^2 + b^2 + 2a \sin \theta \tan \theta + \sin^2 \theta \tan^2 \theta} \qquad (3.27)$$

This method has been used for a number of determinations in the infra-red for surfaces of PbS, PbTe, Ge. The method is precise to about 4 per cent of the value of the absorption index.

4. The previous methods are not applicable to non-isotropic materials such as uni-axial and bi-axial crystals but T. S. Robinson (1952) has evolved a method that is not limited in this way. It is necessary to determine, for normal incidence only, the reflectance for a range of wavelengths. With this method, too, the final result is obtained graphically.

5. **Attenuated (or frustrated) total reflection** (abbreviated ATR or FTR). It follows from expression (3.24) that for $n'/n<1$, total reflection ($R=1$) occurs when the angle of incidence θ is equal to or greater than a ' critical angle '. This is illustrated by the top curve of Fig. 3.10, which shows that for $\theta = 48°$, total reflection occurs when $n'/n \leqslant 0.743$; the top curve,

calculated from expressions (3.22) and (3.24), also indicates for a pair of transparent media the rapid but not 'instantaneous' increase of reflection towards unity as the critical angle or critical refractive index ratio is approached. If the second medium is absorbing, then n' must be replaced by the complex expression (3.26). Calculations based on a combination of (3.22) and (3.26) yield the family of curves in Fig. 3.10. The

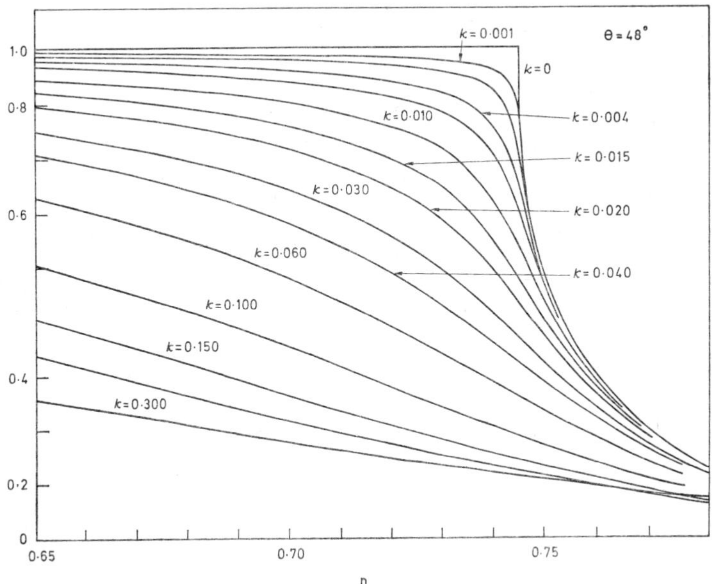

FIG. 3.10. Reflection R related to relative refractive index n for various values of absorption index: all at 48° angle of incidence. κ on the curve is equal to nk, where n = refractive index and k = absorption index of specimen as defined in Table B (Reproduced from Fahrenfort, J., and Visser, W. M., 1962).

reduced or attenuated total reflection indicated by these curves may be explained in physical terms by supposing that, even when there is no refracted beam, the wave penetrates a small but finite distance into the second medium (see dotted ray of Fig. 3.11a). This supposition is substantiated by both theory and experiment [see, for example, page 475 of Longhurst (1967)].

In 1961, Fahrenfort at Amsterdam [see Fahrenfort (1961)

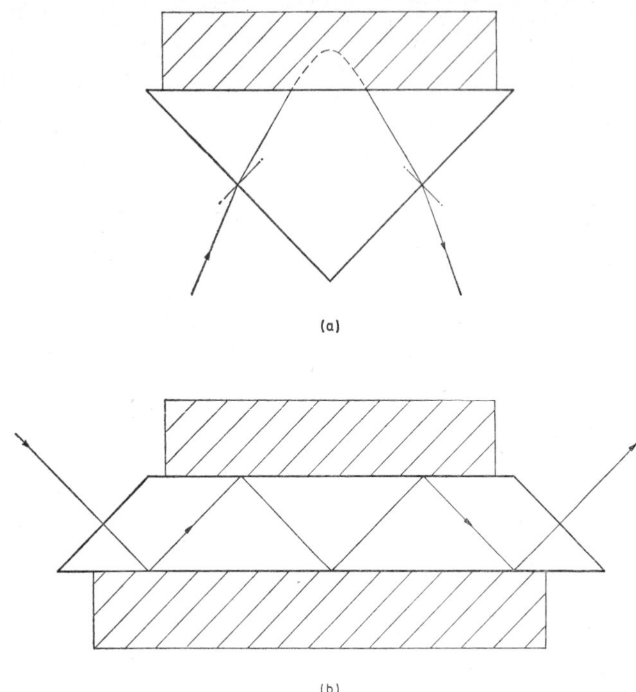

FIG. 3.11. ATR specimen mounts: (*a*) single reflection; (*b*) multiple reflection. The dotted ray is drawn to illustrate penetration into the absorbing specimen.

and Fahrenfort and Visser (1962)] showed how this phenomenon could be developed into a powerful method for the spectrophotometry of materials that cannot easily be obtained as convenient films or solutions. As seen in Fig. 3.11, the specimen to be measured is pressed against the polished face of a prism of material of higher refractive index. The orientation of the prism can be changed so that the angle of incidence on the dividing surface can be varied. Weaker absorption bands can be detected or measured by using a beam that has suffered multiple attenuated reflection, as in Fig. 3.11*b*. As will be explained shortly, it helps if the refractive index of the transparent prism is *considerably* higher than that of the specimen.

In the infra-red, where ATR has particular use, suitable prism materials in order of decreasing usefulness are as follows:

	refractive index	*long-wave limit (transmission 60% for 1 cm)*
thallium bromide-iodide (KRS5)	2·4	40 μm
silver chloride	2·0	14
zinc sulphide (Irtran-II)	2·2	10
germanium	4·0	$\sim$20

Hansen and Horton (1964) discuss cell designs and materials for the use of ATR in the visible and ultra-violet regions.

Hansen (1965) has extended the theoretical aspects of the method. The conclusions of Fahrenfort and of Hansen on the usefulness and limitations of ATR are summarized below.

(*a*) *Useful range of absorptions.* The direct reflection methods (1–4 on pages 43 and 44) are useful for very large absorptions, corresponding to the values exhibited by metals. But ATR can be used for absorptions some 200 times less (see Table 3.1 and Figs. 3.10 and 3.13) and still weaker absorptions can be covered by using a multiple reflection cell.

TABLE 3.1. Useful absorption ranges

Method	Absorption index value, k	Specific extinction coefficient, K per cm
Direct transmission measurement	<0·01	<100
Direct reflection methods (1–4, pages 43–44)	>0·2	>2000
ATR (single reflection)	0·001–0·3	10–3000

(*b*) *Refractive index distortion.* In passing through an intense absorption band the refractive index changes in the manner indicated in Fig. 3.12*b* (anomalous dispersion). This means that the variation of reflection with absorption does not follow a vertical line in Fig. 3.10 and the curve obtained by the ATR method can be quite different in shape from the absorption band (compare Figs. 3.12*a* and *c*); the reflection minimum

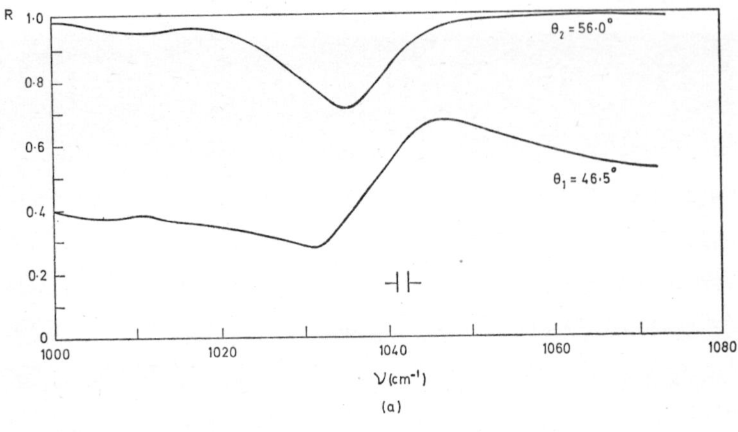

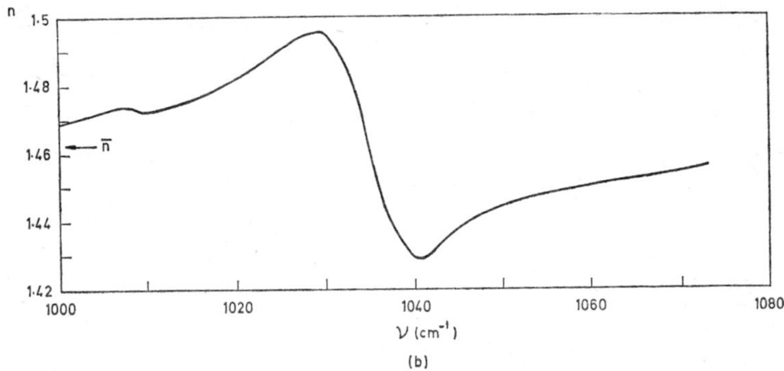

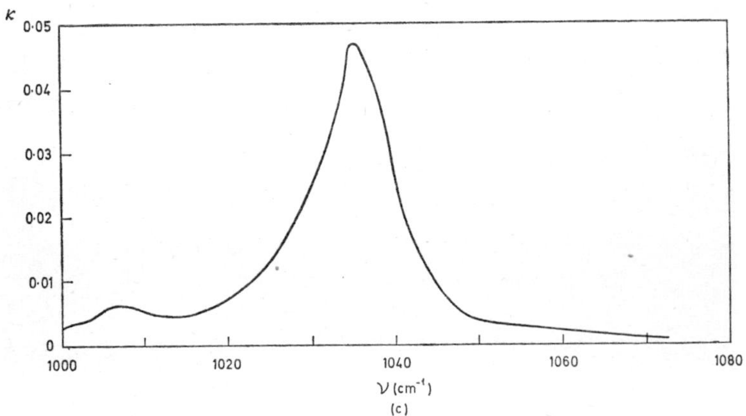

FIG. 3.12. The absorption band of liquid benzene at 1035 cm^{-1}. (*a*) ATR reflection against wave number. (*b*) Refractive index change in passing through the absorption band. (*c*) Change of absorption index in passing through the absorption band.

PLATE 3. Absorption spectrogram of benzene (in hexane) taken with a Spekker ultra-violet photometer and medium quartz spectrograph. Length of tube 2·0 cm. Minimum exposure 5 seconds. The illustration is approximately twice the size of the original spectrogram. The absorption curve constructed from this is seen in Fig. 9.16.

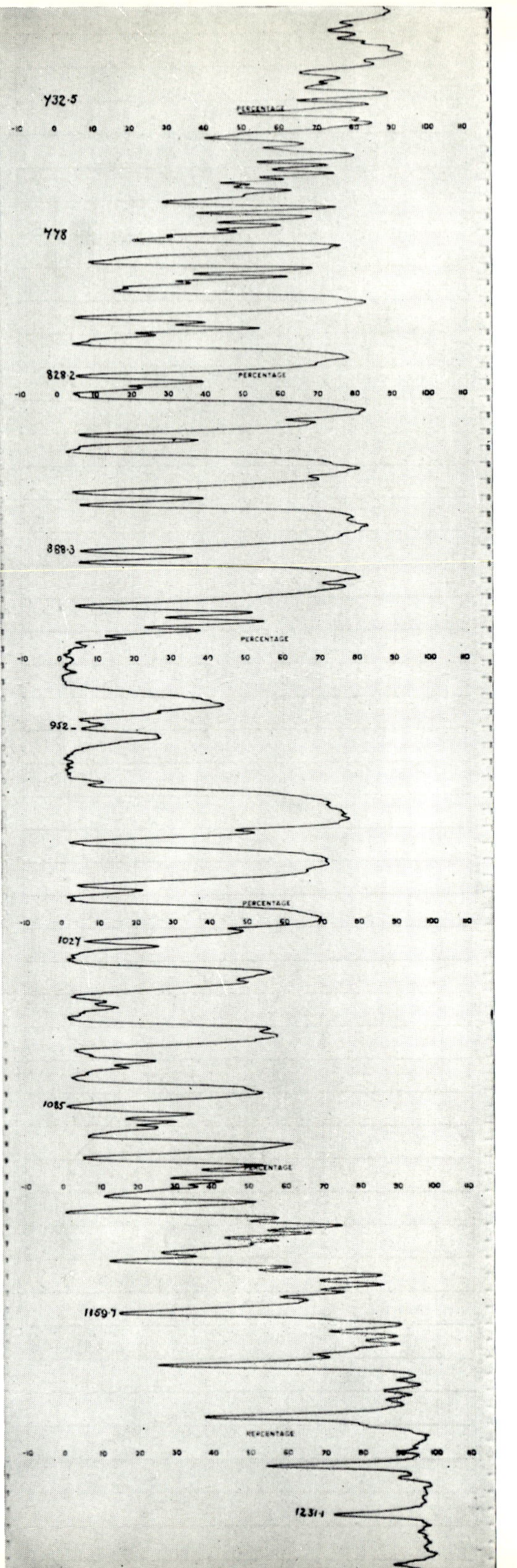

PLATE 4. Double-beam trace of the absorption of ammonia vapour showing the characteristics referred to on page 107 (Hilger H 800 spectrophotometer and Cambridge recorder).

may not correspond with maximum absorption and a spurious reflection maximum may occur (Fig. 3.12*a*, lower curve). This distortion can be reduced by (1) using an angle of incidence further from the critical angle and (2) using a prism material of high refractive index well away from the values for the sample; in this way the variation of *n* through the absorption band is of smaller relative importance.

(*c*) *Intensity distortion.* It is obvious from the complexity of the expressions (3.22–3.27) that there is no simple function of reflection *R* which is proportional to specific extinction co-efficient (i.e., no counterpart to Beer's law); this is implied in the use of graphical treatments in the direct reflection methods 1–4 above. Hansen (1965) has developed the equations to show the relation of $\log(1/R)$—the equivalent of optical

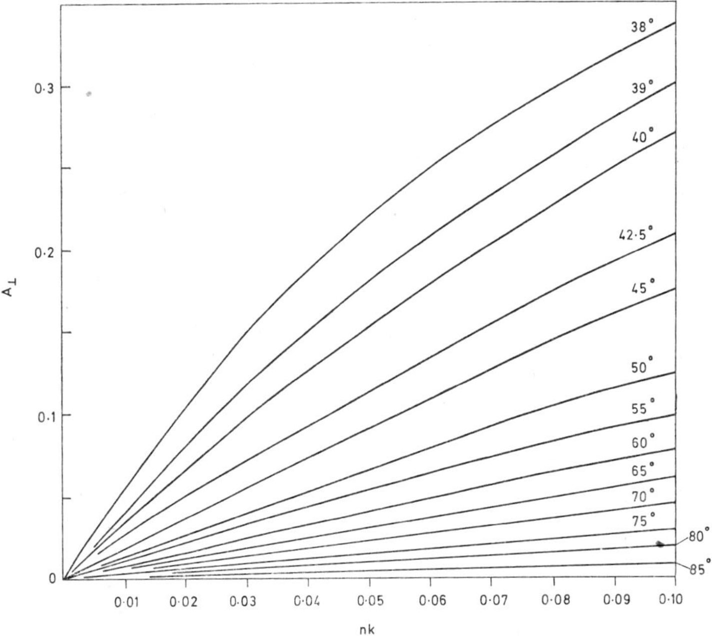

FIG. 3.13. Relation between $A\,(=\log 1/R)$ and nk (refractive index × absorption index for specimen) for plane polarized radiation ($E\perp$), for various angles of incidence at relative index $\mu=0\cdot6$ (Reproduced from Hansen, 1965).

D

density—to absorption index k. While the relation is not linear, which would be the equivalent of Beer's law, it is approximately so for small absorptions (see Fig. 3.13). Similar but not identical curves are shown in his paper for various polarizations of the incident beam.

(d) *Practical use of the ATR method; qualitative and quantitative.* In addition to the distortions mentioned above, a further difficulty may arise in obtaining, except with liquid specimens, a reproducible contact between specimen and prism face. For these reasons, ATR curves are sometimes used for qualitative work only, i.e. as indications of the presence of certain absorption

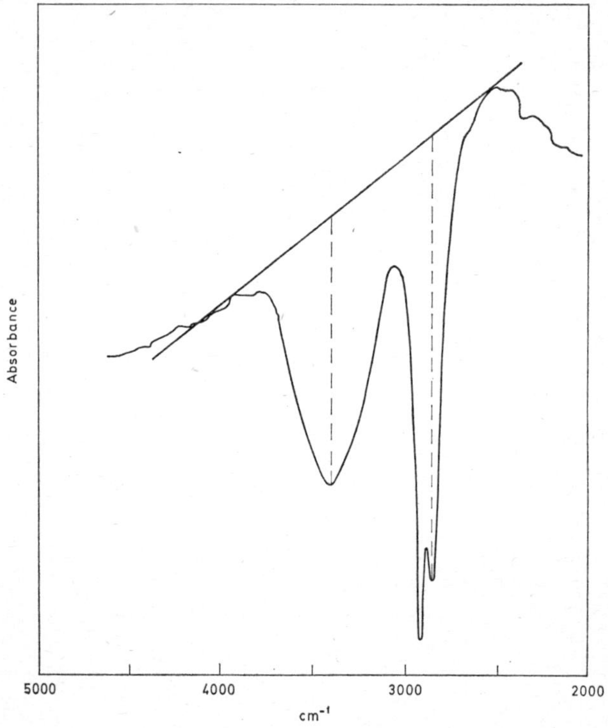

FIG. 3.14. ATR spectrum of a soap specimen pressed against an Irtran-II prism. The absorptions at 2800 and 2910 cm^{-1} are due to the alkyl chain of the fatty acid salt, and the absorption at 3410 cm^{-1} arises from the water content. Ordinate shows ' absorbance ' i.e. log $(1/R)$. (Puttnam, Baxter, Lee, and Stott, 1966.)

bands and hence of certain molecular groups.

But complete quantitative calculations are possible; and measurements at two different angles of incidence can be used to calculate both n and k for a given wavelength; this is more information than can be obtained from direct transmission measurements. Fahrenfort and Visser (1962) have shown how these values can be calculated by use of a computer.

But more often estimations of concentration are made by empirical calibration, as in conventional spectrochemical analysis. This is illustrated in Fig. 3.14 for the estimation of the water content of soap (Puttnam, Baxter, Lee and Stott 1966). The calibration curve in Fig. 3.15 shows the ratio of 'absorbances', i.e. of $\log(1/R)$, for two absorption bands,

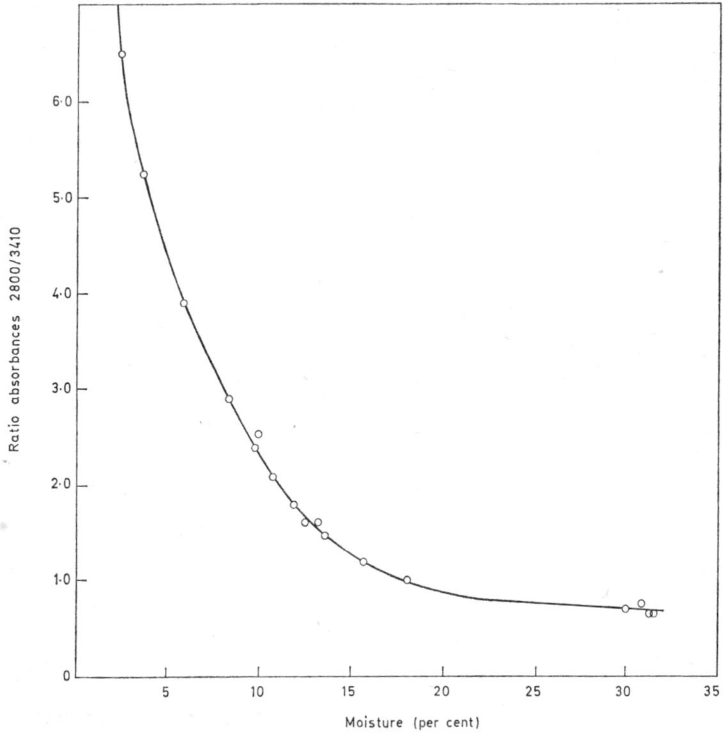

FIG. 3.15. Calibration curve for use with curves of the type shown in Fig. 3.14. (Puttnam *et al.*, 1966.)

one due to the water content and one independent of this. This method, corresponding to the internal-standard method of emission spectrochemical analysis, reduces the effect of variable contact area of the sample.

TOTAL INTENSITY OF AN ABSORPTION BAND

The value of the extinction coefficient or absorption coefficient at the peak of a band depends on a number of factors. For example, if the pressure of a gas is increased so that the energy levels of some atoms or molecules are disturbed, absorption will occur at neighbouring frequencies, i.e. the band will be broadened ; consequently fewer molecules will be present in the original undisturbed state and the band peaks will be depressed. But the total area under the band tends to remain constant and the integral of absorption coefficient over a whole band

$$B = \int \mu \, d\bar{\nu} = (1/lc) \int \log_e (I_0/I) . d\bar{\nu} \qquad (3.28)$$

is an important parameter known as the total intensity of a band.

Values of B may be determined by estimating the area under a curve, either by counting squares or by using a planimeter ; or the integration may be performed more readily by means of a mechanical integrator attached to a spectrophotometer, such as that outlined by Perkin-Elmer (1955).

The theoretical significance of intensity values is discussed in Chapter 6 ; these values may also become of importance in routine analytical work. As already noted, the value of optical density at the peak of a band will depend on the resolution of the spectrometer as well as on the pressure in the case of a gas. This may mean that measurements made on different instruments, or on the same instrument under different conditions, may differ ; individual and repeated calibration is therefore necessary when making routine analytical measurements. Oswald (1954) has pointed out, however, that total intensity varies much less and that changes in resolution that cause 20 per cent variations in peak absorption may cause only a 3 per cent change in the total intensity of a band ; where means are available for conveniently determining total intensities, these values may therefore become very useful for routine analyses. Ramsay (1952) shows that if an absorption band has a Lorentz shape, i.e. if $A = A_{max}/[(\nu - \nu_0)^2 + b^2]$, it is only necessary to measure enough points on a band to determine A_{max} and the width ($\Delta\nu_{1/2}$) at half A_{max} (i.e. half-value width) to permit calculation of the area. He gives a table showing values of total band intensity for various values of A_{max}, $\Delta\nu_{1/2}$ and slit width expressed as a frequency range.

An incompletely resolved band

Integration of the area under a band that shows some structure will not give a true value for the band intensity if the spectrometer used does not completely resolve the structure. Each point on the curve is determined by the mean intensity transmitted by the spectrometer, whereas to give a correct integration it ought to be determined by the mean absorbance. Wilson and Wells (1946) have explained in an important paper the conditions necessary to obtain correct intensities. They show that the intensity is given by the area under the band extrapolated to zero amount of absorbing material—in the case of a gas, to zero value of the product of pressure and length, i.e.

$$B = \mathrm{Lt}_{pl \to 0} \int \mu \, d\nu \qquad (3.29)$$

provided that :

(1) The incident intensity I_0 does not vary greatly over the width of the slit. This condition may not hold when working with a double-beam instrument in the region of atmospheric absorption (see Chapter 9).

(2) Either (a) the instrumental resolving power is sufficient to measure the band structure completely, or (b) the resolving power does not vary greatly over the band.

Uniform resolving power over the band is not, however, of much use unless the band structure can be completely measured, because extrapolation would only be possible by making measurements to very low values of $p \times l$, where the absorbances would often be too small to be measured accurately. It is possible to fulfil condition 2(a) without using an instrument of high resolution by adding

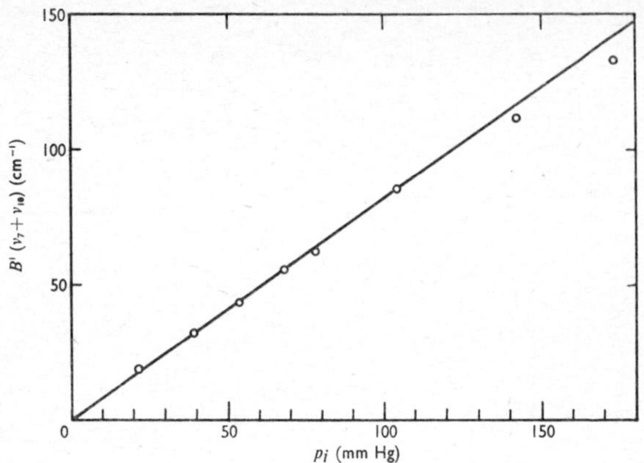

Fig. 3.16. Total intensity of a pair of overlapping bands near $\bar{v} = 1400\ cm^{-1}$ of dimethyl acetylene, at various pressures p_i, mixed with dry air to a constant total pressure of one atmosphere. Ordinates B' are equal to areas under absorption curves, i.e. $B' = \int \log_e(I_0/I)d\bar{v}$. (Mills and Thompson, 1955.)

a non-absorbing gas to broaden the bands so that the structure disappears. This is the normal procedure and its use is illustrated in Fig. 3.16. The linearity of the points at the lower part of the curve shows that the broadening is sufficient, and the gradient of this portion gives the limiting value required by equation (3.29).

Vincent-Geisse (1955) has shown theoretically that the total intensity of a line narrow enough to fall entirely within the mono-chromator slit may be determined from measurements of trans-mission t for decreasing path lengths, using the relation :

$$A = \mathrm{Lt}_{l \to 0}\ a(1 - t)/l$$

where a is the slit width expressed in terms of Δv or $\Delta \bar{v}$.

MEASUREMENTS AT LOW TEMPERATURES

At low temperatures, the interchange of energy between neigh-bouring molecules is reduced, and absorption bands measured at, say, liquid air temperature, are in some cases sharpened to approx-imate more closely to those of a vapour. Fig. 3.17 illustrates this effect for trans-stilbene and in this instance there is a shift of the band at low temperature as well as sharpening. Similar sharpen-ing was observed for liquid benzene and its derivatives by Mayneord and Roe (1937).

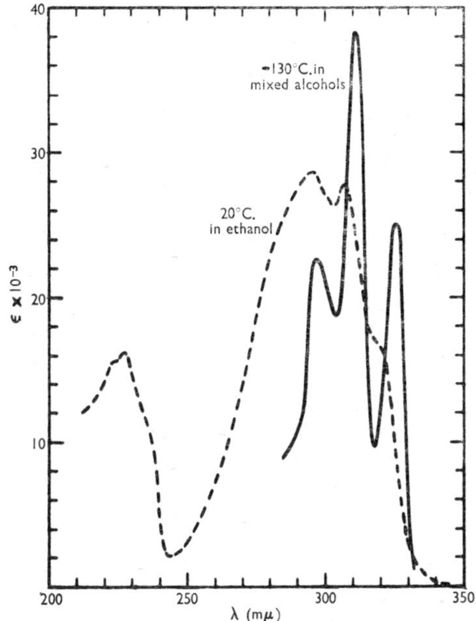

FIG. 3.17. Absorption of trans-stilbene at − 130° C in mixed alcohol, and at +20° C in ethanol. (From Beale, 1951.)

The usefulness of low-temperature measurements became clear when Bowden and Morris (1934) were able to show that irradiated carotene is not identical with vitamin A, as the spectra of the two substances at ordinary temperatures had seemed to suggest. Keilin and Hartree (1949) have observed the spectra of a number of biological materials at liquid air temperatures. In addition to band sharpening, they find an increase of intensity (up to seven-fold) when some materials are cooled in liquid air until the specimen becomes embedded in a mass of microcrystals of ice or other immersing material. This phenomenon is useful when observing weak bands ; it is attributed to the increased path through the absorbing material caused by multiple reflection between the microcrystals. It is noticed that spores of *Bacillus subtilis*, whose cytochrome content is very low, show no evidence of the green-red bands of this substance until the specimen is cooled in liquid air.

In the infra-red, W. H. Avery and Morrison (1947) have observed the spectra of some hydrocarbons at the temperature of liquid air. They find the band widths to be only one-half the widths found at ordinary temperatures, and so there is a much greater chance at low temperatures of finding non-overlapping bands suitable for analysis.

FLUORESCENCE INTENSITY AND CONCENTRATION

Where some fraction of any radiation absorbed is re-emitted as fluorescence, then it may be assumed that the intensity of fluorescence F is proportional to the radiation absorbed. Hence, using the nomenclature of equations (3.6) and (3.7), one may write, where D is a constant,

$$F = D(I_0 - I) \tag{3.30}$$

$$= DI_0(1 - 10^{-acl})$$

$$= DI_0[2 \cdot 3acl - (2 \cdot 3acl)^2/2 + \ldots] \tag{3.31}$$

If c is sufficiently small for the second and further terms in the square bracket to be negligible, then this last expression becomes

$$F = 2 \cdot 3 DI_0 acl$$

i.e. the fluorescence intensity is proportional to concentration. For this to be true to a precision of 1 per cent, the second term should be not more than 1 per cent of the first, i.e.

$$1 \cdot 15 acl \leqslant 0 \cdot 01$$

or, the absorbance should be not more than 0·008. For higher concentrations, the fluorescence intensity will increase more slowly with concentration, approaching asymptotically a limiting value when effectively all the exciting radiation is absorbed.

Quenching effect of absorbing impurities on fluorescence intensity

Equation (3.30) assumes that the fluorescent material is the only one absorbing the incident radiation. But if a solution has extinction coefficients K_1, K_2 for a fluorescent material and for non-fluorescent ' impurities ' respectively, then the total extinction coefficient is $K_1 + K_2$. But in each and every element of the solution only a fraction $K_1/(K_1+K_2)$ is absorbed by the fluorescent component, and equation (3.30) must be replaced by

$$F = D\left(\frac{K_1}{K_1+K_2}\right)(I_0 - I) \qquad (3.32)$$

and equation (3.31) now becomes

$$F = 2\cdot 3 D I_0\left[K_1 l - \frac{2\cdot 3 K_1(K_1+K_2)l^2}{2} + \cdots\right] \qquad (3.33)$$

From this last equation we conclude that, if the absorptions of the fluorescent material and the impurity are both small enough to neglect the second and higher terms, then the fluorescence is (a) proportional to the concentration of material 1 and (b) independent of the impurity concentration. At higher impurity concentrations, there will be partial 'quenching' of the fluorescence. But even if such higher terms are negligible, another form of quenching may operate—the 'impurity' molecules may, by collision with the excited fluorescent molecules, deactivate them before they have time to emit the fluorescent radiation.

Equations (3.31) and (3.33) are illustrated by some measurements (Lothian, 1941) on dilutions of a solution of uranine with and without 2-naphthol as an absorbing impurity:

Conc. of uranine 75 µg/cc	×1	1/2	1/4	1/8	1/16	1/32	1/64
Fluorescence:							
(a) Pure uranine	100·0	76·8	45·0	24·0	12·3	6·1$_7$	3·1
(b) Uranine plus 2-naphthol	63·1	—	42·7	—	12·4	6·2	3·1
Effective absorbance of uranine solution	0·82	0·41	0·20	0·10	0·05	0·02$_5$	0·01$_2$

Quenching is a prolific source of error in fluorimetric estimations; but its effect may be determined and allowed for in the following way. One determines the changes of fluorescence F_1, F_2, on adding equal quantities of the material estimated to solvent (blank) and test solution respectively. The ratio F_1/F_2 gives the factor by which the measured fluorescence must be increased to correct for the effect of absorbing impurities. Booth (1940) and Hodson and Norris (1939) have used this method for estimating vitamin B_1. The latter workers say that the method is satisfactory even when impurities are present to such an extent that the fluorescence is reduced to one-fifth of its normal intensity.

CHAPTER 4

Accuracy in Spectrophotometry

CLASSES OF SPECTROPHOTOMETRIC MEASUREMENT

Before describing the various instruments used for spectrophotometry it will be useful to consider in a general way the forms of design that will give the greatest accuracy for a particular purpose.

Spectrophotometric measurements can be classified according to the nature of the problem to which they are to be applied. Possible applications tend to fall into two main categories—those in which we seek information on the nature and state of the absorbing medium and those in which we are interested in the effect produced by the medium on the radiation transmitted or reflected by it. Thus we may measure the absorption in order to identify and estimate the absorbing substances, to decide some question of molecular constitution, or to ascertain what is taking place inside the atoms or molecules—in each instance it is the proportion of radiation *absorbed* that we desire to ascertain.

In the first category, the function of absorption in terms of which it is appropriate to express results is the extinction coefficient K, which, for any substance and wavelength, is proportional to the concentration of the absorbing material (Beer's law). Thus the proportional accuracy with which a concentration may be determined is $\Delta c/c$ and by differentiation of equation (3.6) it is seen that

$$\Delta c/c = \Delta A/A \qquad (4.1)$$

assuming that the cell length and specific extinction coefficient are accurately known or that the absolute values of these are not involved, as in the comparison of two solutions of the same substance in the same cell.

On the other hand, we may use analogous methods to test media such as coloured filters and fabrics, and the glasses with which spectacles are made, in order to ascertain their effect on the *transmitted* or *reflected* intensity I. The proportional error of its measurment is represented by $\Delta I/I$, where ΔI is the smallest measurable increment of I, the intensity of the transmitted or reflected radiation.

The two classes of measurement require differing methods. Because this book is primarily concerned with the first class, the emphasis in this chapter is on methods applicable to it.

ABSORPTION MEASUREMENTS

The various constructional features and operating conditions affecting the use of a spectrophotometer for absorption measurements may be discussed under three main headings :

1. The dispersing system (including effects of stray radiation).
2. Cells for liquids and gases.
3. Photometric methods.

THE DISPERSING SYSTEM

The points of an absorption curve where great accuracy is required will often be on the slope, so that, if the measurements made on different occasions are to correspond, we must ensure that the wavelengths used are always the same, within the limit of the following condition—that the error of absorbance due to error in wavelength setting should be less than the error in the photometric measurement of the absorbance. An instrument of sufficiently high dispersion is therefore needed when making use of a continuous spectrum ; the steeper the curve at the point under consideration the greater the dispersion desirable.

This is illustrated in the use of a spectrophotometer for determining small percentages of carboxyhaemoglobin in blood, when useful accuracy can be attained by making measurements near wavelengths of 5600 Å and 5770 Å (see the absorption curves in Fig 7.6 on p. 146). At the latter wavelength the spectrophotometric curve of carboxyhaemoglobin is so steep that very narrow slits must be used in order to get a significant spectrophotometric measurement.

In consequence of the low intensity of the light transmitted at large optical densities, it may sometimes be desirable to increase the width of the spectrometer slits. When the source has a line spectrum, the slits can be opened considerably without sacrificing purity of spectrum, because the same lines can then be used on each occasion and the condition stipulated above for wavelength setting is automatically fulfilled. With a continuous-spectrum source, however, it is necessary to compromise between the best light intensity and the desired purity of spectrum.

Important considerations governing the design of the dispersing system are :

1. The intensity of 'monochromatic' radiation required.
2. The waveband or wavelength range included in the mono-chromatic beam.
3. The relative amount of stray radiation of other wavelengths.

The intensity and waveband of a beam passed by a mono-chromator depend in turn on :

(*a*) The intrinsic brightness of the light source.
(*b*) The slit sizes.
(*c*) The focal length of the system.
(*d*) The size and dispersion of the prism.

Suppose that, at wavelength λ, the intrinsic intensity of a light source, i.e. the intensity emitted per unit area per unit solid angle per unit wavelength range, is I_0; for a wavelength range $\delta\lambda$ this may then be written $I_0\delta\lambda$. If a light source be then imaged on the entrance slit, and reflection losses ignored, the same quantity $I_0\,\delta\lambda$ is radiated per unit solid angle from unit area of slit.

Suppose that the width and height of the entrance slit are x and y respectively, that A is the area of beam that fills the prism, and that f is the focal length of the collimator lens. Neglecting diffraction effects (see below) and reflection losses, the energy passing through the slit and falling on the prism ($I=I_0\,\delta\lambda \times$ slit area $\times$ solid angle subtended by the prism at the slit) may then be expressed by $I=I_0\,\delta\lambda\,.\,xy\,.\,A/f^2$. If the angular dispersion of the prism is $d\theta/d\lambda$, then the wavelength range included in the entrance slit is

$$\delta\lambda = \frac{d\lambda}{d\theta}\cdot\frac{x}{f} \tag{4.2}$$

Combining this with the previous equation to eliminate x,

$$I = I_0\,\delta\lambda \cdot y \cdot (A/f^2) \cdot f(d\theta/d\lambda) \cdot \delta\lambda$$

and therefore

$$I = I_0(y/f) \cdot A \cdot (d\theta/d\lambda)(\delta\lambda)^2 \qquad\qquad (4.3)$$

From this we see that the intensity is proportional to :

(*a*) The intrinsic intensity of the source.

(*b*) The dispersion of the prism.

(*c*) The square of the wavelength range. (This is obvious when we remember that to double the waveband we can double both entrance and exit slits.)

(*d*) The aperture of the prism—assuming that the lenses are so large that the whole of the prism aperture is filled with radiation.

(*e*) The ratio of slit height to focal length.

Thus the light intensity is independent of focal length so long as y/f is kept constant. Reasons for designing an instrument with long focal length are that it is then possible to get greater optical perfection (freedom from aberration) and to use wider slits for the same waveband, with less demand for high mechanical accuracy in the slits. In the case of a spectrograph, a larger spectrogram is obtained ; this is important because the resolving power might otherwise be limited not by optical conditions but by the grain size of the photographic plate. On the other hand, a short focal length means a more compact instrument, and compromise is necessary. In practice, intensity is reduced by reflection losses at the various surfaces and the greater the number of surfaces the greater the total loss. The transmission problems associated with a monochromator are dealt with very fully by Perry (1938 and 1956), by Bauer (1956), and by Conn (1956).

Diffraction effects and resolving power

The wavelength range passed by a given slit is greater than indicated by equation (4.2) on account of diffraction effects produced by the finite aperture of the prism. It is shown in text-books of optics that, for a beam of width a at the prism and for perfectly monochromatic light of wavelength λ, the image at the exit slit is wider than the geometrical image of the entrance slit by an angular amount on each side

$$\delta\theta = \frac{\lambda}{a} \qquad (4.4)$$

With a continuous spectrum, the waveband passed when using infinitely narrow slits is therefore given by

$$\delta\lambda = 2\frac{\lambda}{a} \cdot \frac{1}{d\theta/d\lambda} \qquad (4.5)$$

and for a finite exit slit the waveband is this much greater than the geometrical amount.

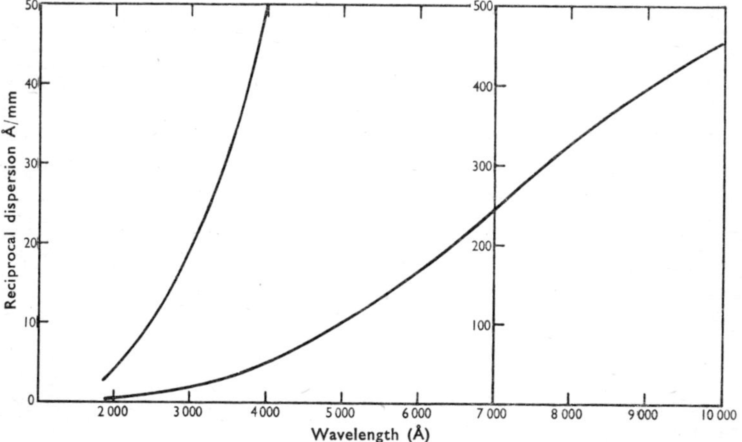

FIG. 4.1. Dispersion of quartz between 2000 and 10,000 Å. For a 60-degree prism and focal length of 100 cm. The left-hand and centre ordinate scales refer to the left-hand and right-hand curves respectively.

Equation (4.5) clearly determines the least change of wavelength that can be resolved by a spectrometer. According to the convention adopted by Lord Rayleigh, the smallest wavelength change that can be resolved by an optical system is equal to this expression, but such an interpretation is fairly lenient ; it is sometimes found that with a receiver sensitive to small changes of intensity, such as a thermopile, resolving power is slightly better than that given by the Rayleigh expression. This is most likely to apply in the infra-red, where scattering effects are much smaller than elsewhere.

Equation (4.5) shows that the resolving power depends on the wavelength, prism size and prism material (dispersion). The numerical values for a 60-degree prism using a beam width of about 6·5 cm at a wavelength of 7 μ give a useful example ; with a rocksalt prism the Rayleigh resolution in wavenumbers is 4·5 cm^{-1} but for a fluorite prism, which has greater dispersion, the resolution is 1·1 cm^{-1}.

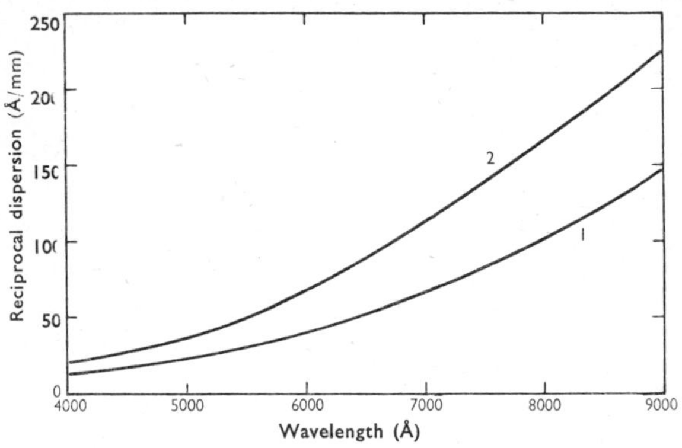

FIG. 4.2.   Dispersion of glass.   (1) Extra-dense flint, $n_D = 1·74$.   (2) Dense flint, $n_D = 1·62$.   Curves for 60-degree prism and focal length of 100 cm.

Choice of prism

Equation (4.3) shows that a prism material of high dispersion is required for high intensity and equation (4.5) shows that high dispersion is also necessary if the increase of waveband due to diffraction is to be small.   From this it may be concluded that a prism material giving highest dispersion for the wavelength in question is to be preferred.   Figures 4.1, 4.2 and 4.3 show the linear dispersions $d\lambda/dx = (1/f)\, d\lambda/d\theta$ for a number of 60-degree prisms for a focal length of 100 cm.   For other focal lengths the dispersions can be found by applying simple proportions.   When working, for example, at say 6000 Å in the visible, the use of an extra-dense-flint glass gives a waveband about a fifth of that given by a quartz prism under similar conditions.

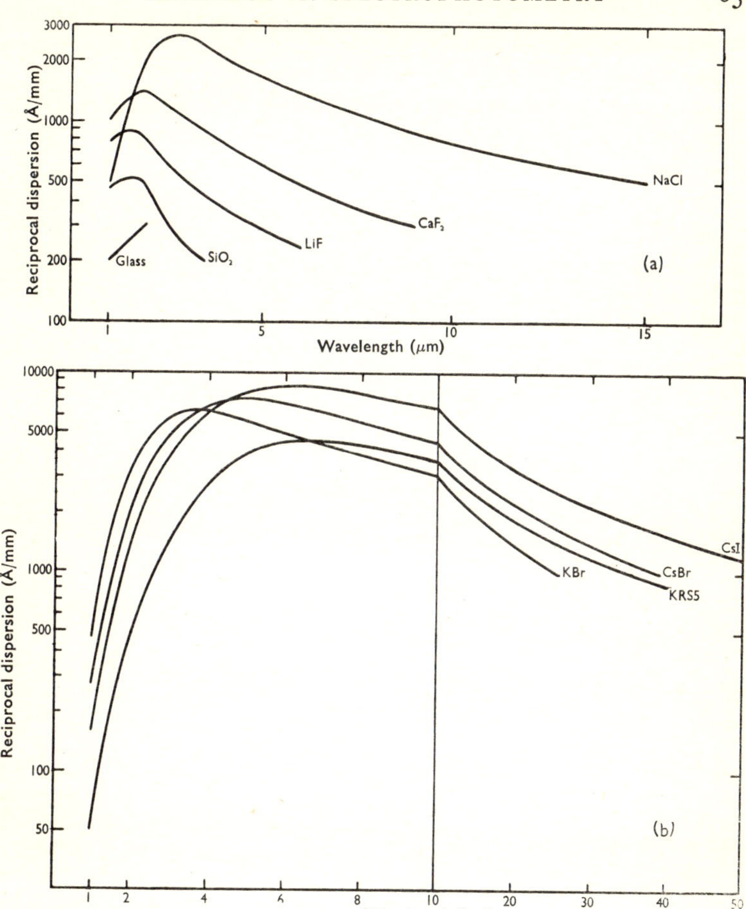

FIG. 4.3. Dispersion curves in the infra-red. (a) Glass, quartz, lithium fluoride, calcium fluoride, and sodium chloride. (b) Potassium bromide, caesium bromide, caesium iodide, and thallium bromo-iodide (KRS5). Curves are for a focal length of 100 cm and for a 60-degree prism (with thallium bromo-iodide, two 30-degree prisms are assumed since total reflection rules out a single 60-degree prism).

Figs. 4.4 and 4.5 show the approximate extinction coefficients of various materials, ignoring losses due to reflection at the surfaces. Transmissions for different thicknesses may be calculated by applying Lambert's law. At wavelengths for which the transmission is decreasing, only the apex of the prism is transmitting

E

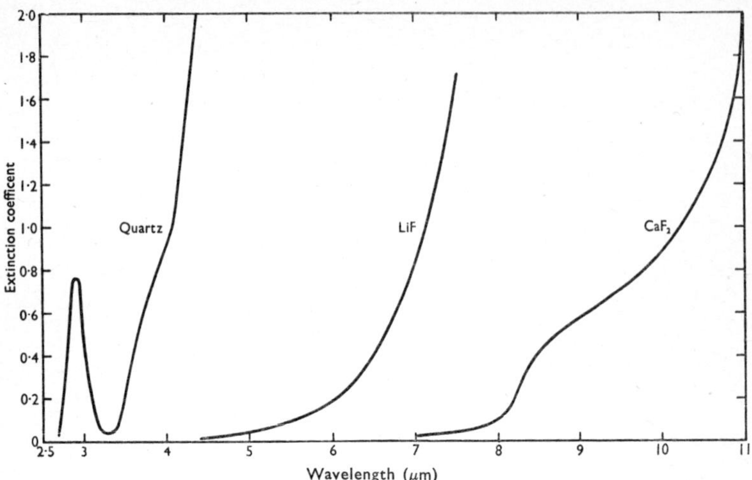

FIG. 4.4. Absorption curves for quartz, lithium fluoride, and calcium fluoride. Ordinates are extinction coefficients per cm.

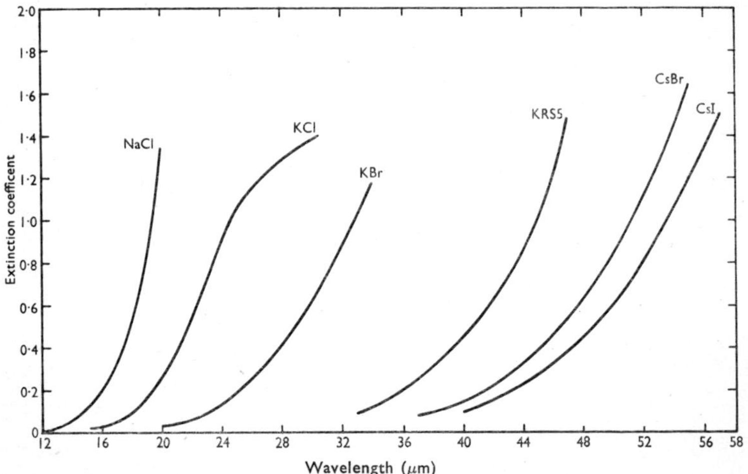

FIG. 4.5. Absorption curves for sodium chloride, potassium chloride, potassium bromide, thallium bromo-iodide (KRS5), caesium bromide, and caesium iodide. Ordinates are extinction coefficients per cm.

effectively and the part of the prism near the base is not being utilized. This means a loss not only of light intensity but also of resolving power, the effective beam width being less than for the full prism.

At a wavelength at which the transmission of a prism is small, a thin window of the same material may transmit quite adequately. At $\lambda = 4\,\mu$, for example, the transmission of a quartz prism is negligible, and a prism of rocksalt or some other material is necessary. Thin quartz cell windows, however, may still be used with some gain in convenience because, unlike rocksalt, they do not need protection from the atmosphere.

Stray radiation

In practice, the exit slit of a monochromator will pass not only the narrow waveband centred on the wavelength drum setting but also some radiation of other wavelengths scattered in the monochromator. This scattered radiation, also known as stray or background radiation, will generally be only a small fraction of the wanted radiation; in certain limiting circumstances described below it may, however, become large and even approach 100 per cent. The effect of this stray radiation will be to give incorrect absorbance values. Thus if I_0 and I are the intensities of the incident and transmitted beam of the 'wanted' radiation, and S is the intensity of stray radiation supposed to be completely transmitted by the specimen, the measured absorbance will be $\log_{10}[(I_0+S)/(I+S)]$. For high absorbance (I small compared with S) the stray radiation S will have a preponderating effect and, if the concentration of a solution is increased, the absorbance, instead of increasing towards infinity, will approach a maximum value

$$A_{\max} = \log_{10} \frac{I_0 + S}{S} \tag{4.6}$$

so that the curve of measured absorbance against path length is not a straight line, but curves over to become parallel with the axis denoting path length (see Fig. 3.1b, p. 26).

The shape of such a curve may be used to measure the stray radiation S as a fraction of I_0—the theory is set out in a careful paper by Hogness, Zscheile and Sidwell (1937). Such a procedure may need to be carried out when the amount of stray radiation is thought or known to be appreciable. Thus the above authors found, for an ultra-violet monochromator using a hydrogen lamp as source, that at 3700 Å (with a potassium chromate solution, which has a peak of absorption at this wavelength) the stray is 0·4 per cent of the wanted radiation and at 2200 Å it is 1·1 per cent. Pritchard (1955) has determined the stray radiation of a mono-chromator by feeding into it pure monochromatic radiation from a double monochromator and measuring the output for various wavelength settings of the instrument under test. The process is repeated for various input wavelengths and the results expressed by a square array of terms or by a series of curves. It is shown how the values obtained may be used to estimate the correction necessary for an absorbance measurement made under given conditions.

There are several ways of designing and using a spectrophoto-meter with a view to reducing the effects of stray radiation :

1. *General design*. The stray radiation will be decreased by proper attention, when designing an instrument, to elimination of unwanted reflections, and constant care later to remove dust and keep all optical surfaces clean. ' Blooming ' the optical surfaces to reduce their reflection is now routine in some monochromators.

2. *Double monochromator*. Two monochromators may be used in series, forming what is known as a double monochromator, the exit slit of the first serving as the entrance slit of the second. Stray radiation transmitted through this slit is dispersed in the second monochromator and only a very small fraction of it emerges through the final slit. Another advantage of the double mono-chromator is that the dispersion may be twice that of each half, giving increased resolution and light-gathering power.

3. *Multi-pass monochromator*. The double monochromator, with its duplication of optical components, is extravagant in cost and space. A. Walsh (1951 and 1952) has developed a multi-pass monochromator in which the beam is made to traverse the same monochromator several times. It has the advantages of the double monochromator without its disadvantages. In the arrangement illustrated in Fig. 4.6, radiation of a certain wavelength arrives near the exit slit S_2 at the mirrors M_3 and M_4 and is reflected back to pass again through the Littrow monochromator, whence it is

finally returned to the exit slit. Radiation of a different wavelength that has passed through the system once only (first-order radiation), will also emerge through S_2 but the second-order radiation is interrupted by a chopper C near M_4 and if the detector

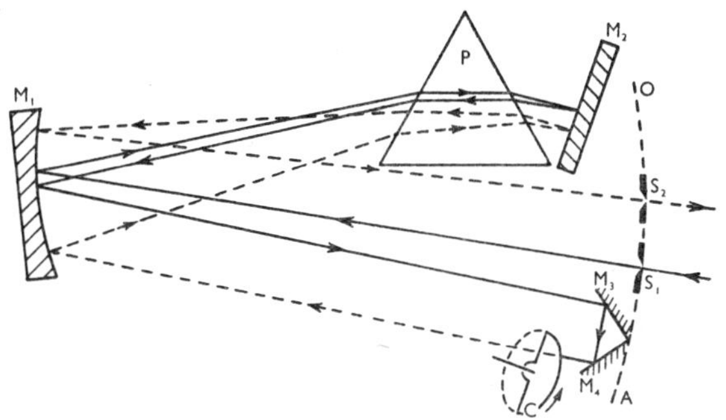

Fig. 4.6. Walsh double-pass monochromator. (From Walsh, 1952.)

includes a tuned amplifier responding to the frequency of chopping the unchopped first-order radiation will not be detected. Walsh (1952) states that, for a Perkin-Elmer 12C spectrometer modified by him in this way, ' beyond 18 μm, using a KBr prism, the stray radiation transmitted through a rocksalt shutter 10 mm thick is too small (less than 0·25 per cent) to be detected'. The principle may obviously be extended to pass the radiation through the monochromator several times and Ham, Walsh and Willis (1952) have built a quadruple monochromator.

4. *Littrow monochromator.* In a Littrow monochromator (Fig. 4.6) much of the stray radiation is caused by reflections from the prism surfaces. If the radiation reflected by the Littrow mirror is chopped, either by means of a rotating sector in front of it (Walsh, 1953) or by rocking the Littrow mirror about an axis perpendicular to the refracting edge of the prism (Hammond and Price, 1953), the unchopped stray radiation will be undetected by a tuned amplifier.

5. *Filters.* If a suitable optical filter, chosen to transmit the required wavelength and to absorb as much as possible of other wavelengths, is placed in the beam, it forms a crude mono-

chromator and is a much used device for removing stray radiation. A number of suitable filters are described in Table 9.1 on page 190, others may be chosen with the help of Table 10.1 on page 210.

6. *The detector.* The effect of stray radiation is determined also by the nature of the receiver of the radiation—which may be the eye, a photocell, a thermopile, etc. When the receiver is selective, responding to certain wavelengths only, the effect may be helpful or disturbing. Let us consider, for example, the use of a visual spectrophotometer where the eye is the receiver. If one were using monochromatic radiation at $\lambda = 5500$ Å in the green, where the eye has its maximum sensitivity, small amounts of blue and red scattered light would have only a small effect. But working at, say, 4000 Å, where the sensitivity of the eye is very limited, a small amount of stray green radiation could completely mask the violet light of 4000 Å. Circumstances are similar when a photocell or photographic plate is used at wavelengths where its sensitivity is small. In general, short-wave radiation is scattered more than long-wave and there is therefore likely to be more stray radiation in the short-wave ultra-violet than in the visible and infra-red. Near the limits of transmission (near 2000 and 30,000 Å for a quartz and near its 15-μm limit for a rocksalt monochromator) appreciable wanted radiation is absorbed and the proportion of unwanted radiation is increased.

7. *The light source.* The greater the proportion of wanted radiation to the total radiation from the source, the less will be the effective stray radiation. Thus, if a monochromator is used to isolate the green line from mercury radiation where this line constitutes perhaps 20 per cent of the total radiation, the purity will be much greater than if a band of narrow wavelength near 5461 Å is isolated by the same monochromator from tungsten lamp radiation, when the required waveband may constitute only a very small fraction of the total energy—about 0·2 per cent from a waveband 10 Å wide for the visible radiation from a source of colour temperature 2848° C. An example important in infra-red spectroscopy is a Nernst filament ; it has maximum energy near 1 μm and only a small fraction of this maximum is emitted at infra-red wavelengths beyond, say, 6 μm, so that stray radiation is likely to

be important in the latter region. The stray radiation from a monochromator may well vary during its lifetime, depending on the cleanliness and adjustment of its components, and a test will be useful in indicating the condition of the instrument.

CELLS FOR LIQUIDS AND GASES

The accuracy of the cells should be such that the departure from the nominal length can cause no appreciable error.

Photoelectric measurements may be made with great accuracy and, with the transmission-ratio method outlined on page 78, may be of as good precision as 0·04 per cent. It is probable that cells will not be interchangeable to this degree of precision and it will be necessary for the user to compare individual cells. This may be done by measuring them when filled with a standard solution of optical density A_s, permitting high accuracy of measurement, and then when filled with solvent A_0. The length of a cell is then proportional to $A_s - A_0$.

Accuracy is required in setting up the cell. Unless the cell is each time set perpendicular to the light beam, the length of the light path through the absorbing medium will vary. Suppose that the normal to the cell makes a small angle i with the incident light beam. If n is the refractive index of the absorbing medium, the fractional error δ in the path is given by

$$i = n\sqrt{(2\delta)} \qquad\qquad (4.7)$$

For instance, if $n = 1\cdot3$ and $\delta = 0\cdot0004$, then $i = 2°$. Thus the cell should be set with its faces perpendicular to the light beam to within two degrees. As the cell can readily be set to within one degree, the error should be negligible, although special care must be taken when setting short cells in their mount. With long cells, 5–10 cm for instance, sideways displacement of the beam must be guarded against because it may make the illumination in the two beams unequal. The length of cell chosen should be such as to give an absorbance approximating to the optimum value for the method used. For this reason, it may sometimes be an advantage to use different cells for different portions of the same absorption band.

PHOTOMETRIC METHODS

We must now consider broadly the means of photometry which
are available and the best way of employing each to determine
extinction coefficients accurately. The means we shall consider
are :

1. Direct measurement by an objective photometer employing
   photoelectric cells, thermopiles or bolometers. Although
   such devices vary considerably, they share certain requirements that make it convenient to discuss them together in
   this chapter on accuracy.
2. Photographic spectrophotometry.
3. Visual spectrophotometry.

Logarithmic differentiation of the expression $A = acl$ [equation
(3.6)] leads to equation (4.1),

$$\Delta c/c = \Delta A/A$$

if we assume that the uncertainties in a and l are negligible or
irrelevant. Thus the smallest fractional error in measuring a
concentration is equal to the smallest fractional value of $\Delta A/A$.
The value of absorbance which gives this minimum error in any
particular set of circumstances needs careful consideration.

1. OBJECTIVE SPECTROPHOTOMETRY

The precision may be limited by one or more of several fundamental sources of fluctuation and by other factors which are
peculiar to particular designs. The first or fundamental group is
generally named noise.

Noise

Fluctuations under this heading arise because of the atomic or
particulate nature of our universe. The fluctuations are related
to the Brownian movement of small particles, the theory of
which was elucidated by Einstein in 1905. The subject of noise
can be treated only very briefly here; it is covered very clearly
in its relation to infra-red spectroscopy by Conn and Avery (1960)
and at rather greater length by Smith, Jones and Chasmar (1957).
The various fundamental sources of noise are classified thus:

1. *Johnson noise.* In any conductor there is a random motion (Brownian motion) of the current carriers (electrons) giving a fluctuating current (Δi) or e.m.f. (ΔE). In accordance with Einstein's theory of Brownian motion, the energy associated with these fluctuations is approximately equal to kT, where k is Boltzmann's constant; i.e.

$$[(\Delta E)^2/R] \cdot t = [(\Delta i)^2 R] \cdot t = kT \tag{4.8}$$

where R is the resistance of the conductor and t is the time over which the square of the fluctuation current or e.m.f. is averaged. When the measurement is made by a galvanometer, t is its response time. In the more common arrangement, in which a chopped radiation beam passes to a detector and an amplifier tuned to the chopping frequency, the t of the above expression should be replaced by $1/\Delta f$, where Δf is the width of the frequency band to which the amplifier responds. More careful analysis introduces a factor of 4, so that (4.8) becomes

$$\Delta E^2 = 4kTR \cdot \Delta f \tag{4.9}$$

2. *Thermal noise.* The temperature of a detector shows random variations because of the discrete units in which heat energy arrives and leaves—by conduction and convection and by radiation (photons). Where radiation interchange alone is concerned it can be shown that

$$\Delta W^2 = 16Ak\sigma\epsilon T^5 \cdot \Delta f \tag{4.9a}$$

where ΔW^2 is the mean square fluctuation in the power (intensity) flow to and from the body, ϵ is the surface emissivity ($\epsilon = 1$ for a black surface), σ is Stefan's constant and A is the surface area.

3. *Photon noise* is a manifestation of the random arrival of photons in a beam of radiation.

4. *Shot noise*, first considered by Schottky in 1918, is the randomness of the thermionic current emission in a valve, due to the finite charge of the current carriers (electrons).

Simple statistical considerations lead to the conclusion, for both photon noise and shot noise, that the mean square fluctuation Δn^2 of the number n of photons or electrons emitted is given by

$$(\Delta n)^2 = n \tag{4.10}$$

For electrons, if the time of observation is t, then the current is $n\,e/t$ and

$$(\Delta i)^2 = (\Delta n \cdot e)^2/t^2 = n\,e^2/t^2 = e\,i/t \qquad (4.11)$$

Where the current in a valve is space-charge limited, the shot noise can be somewhat less than this.

Some other sources of noise, not fundamental in the sense of the above, are as follows.

5. *Current noise.* Current fluctuations (over and above Johnson noise) in conductors (not metals) that are carrying a current. This is not fully understood.

6. *Microphonic noise*, arising from mechanical vibration.

7. *Flicker noise*, which occurs over and above shot noise with oxide-coated thermionic emitters.

Other less fundamental sources of error in spectrophotometers are as follows:

8. Instability in the amplifier following the detector, due, for example, to voltage fluctuations in the power supply feeding the amplifier.

9. In single-beam non-recording instruments (see, for example, Fig. 9.2) fluctuations in the intensity of the source of radiation in the few seconds between the settings on reference standard and test specimen will affect the readings. These fluctuations may arise from variations in the voltage supply to the lamp and by irregular convection effects outside and (for gas-filled lamps) inside the lamp. This leads to a fluctuation Δi in the signal i which is proportional to the signal, i.e.

$$\Delta i/i = \text{constant} \qquad (4.12)$$

10. Closeness of the divisions on the potentiometer scale (on which readings are in some cases obtained) and non-uniformity of resistance of the potentiometer wire.

11. In balanced-beam instruments with a comb or other shape of diaphragm, uniformity of illumination over the area of the aperture is important.

12. Cleanliness and matching properties of cells for holding liquids.

Photoelectric and thermoelectric instruments (ignoring, until later, instruments using photomultipliers).

Primary sources of error are as follows.

(*a*) Johnson noise (number 1, page 73); equation (4.8) shows that Δi is independent of signal i, i.e.

$$\Delta i = \text{constant} \tag{4.13}$$

(*b*) Fluctuations of types 8, 9, 11 and 12 in the above list, for which the signal fluctuation is proportional to the signal, i.e.

$$\Delta i/i = \text{constant} \tag{4.14}$$

We shall consider in turn the consequences of these two factors:

(*a*) $\Delta i = \text{constant}$. Remembering that $A = \log (I_0/I)$ and differentiating leads to

$$\Delta A = -\log_{10} e \cdot \Delta I/I$$
$$= -0.43 \Delta I/I_0 \cdot 10^{-A}$$

Hence equation (4.1) becomes

$$\Delta c/c = \Delta A/A = -0.43 \Delta I/I_0 A \cdot 10^{-A} \tag{4.15}$$

Since i is proportional to I, i.e. $\Delta I = \text{constant}$, $\Delta c/c$ is a minimum when $A \cdot 10^{-A}$ in the denominator is a maximum. Differentiation of this last expression shows that it has a maximum value at

$$A_{\text{optimum}} = 0.43 \tag{4.16}$$

Curves b and c of Fig. 4.7 show this minimum error and also that the error is not greatly increased for any absorbance in the range 0.2–0.8.

The expression (4.16) has been deduced by a number of authors, starting with von Halban and Eisenbrand (1927). For many years this was the only factor considered in assessing the error of a spectrophotometer, and exclusion of sources of error under heading (*b*) above has given rise to considerable confusion.

(*b*) Sources of error for which $\Delta i/i = \text{constant}$.

Differentiation of $A = \log (I_0/I)$ shows that this condition leads to

$$\Delta A = \text{constant} \tag{4.17}$$

Comparing this with equation (4.1) indicates a decreasing error for increasing absorbances, as indicated by curve d of Fig. 4.7.

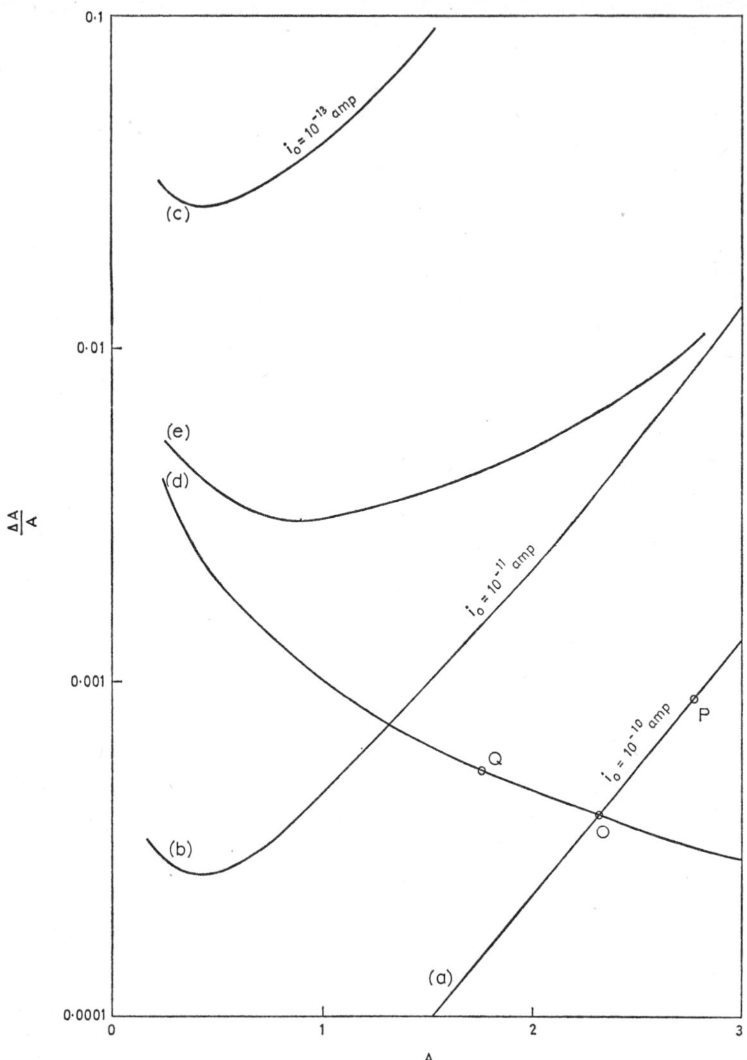

FIG. 4.7. Fractional error $(\Delta A/A)$ in a measured optical density plotted against values of optical density. Curves a, b, c assume $\Delta i =$ constant $= 10^{-15}$ ampere and $i_0 = 10^{-10}$, 10^{-11} and 10^{-13} amperes respectively (equation 4.13). Curve d assumes $\Delta i/i =$ constant (equation 4.14). Curve e assumes $\Delta i/\sqrt{i} =$ constant (equation 4.18).

Practical experience with a number of spectrophotometers indicates that under suitable conditions the constant value of ΔA is about 0·001–0·002.

In any practical case the actual error will be determined by the greater of the two types of error; i.e. at any absorbance the limiting error will correspond to the higher of the two curves a (b or c etc.) and d. This conclusion may be elucidated by considering two specific examples.

1. Considering a single-beam non-recording instrument for which the measuring circuit is fundamentally that of Fig. 8.2 (page 165), we shall suppose that Δi of equation (4.13) is determined only by the Johnson noise of the resistance in series with the photocell; a typical value for this resistance, which is also convenient for calculation, is $R=4·2 \cdot 10^9$ ohms. Using equation (4.8) for Johnson noise and taking the time for a measurement (response time for the output galvanometer) to be one second, we obtain

$$\Delta i = \sqrt{(kT/Rt)} = \sqrt{(1·4 \cdot 10^{-23} \cdot 300/4·2 \cdot 10^9)}$$
$$= 10^{-15} \text{ ampere}$$

(It is probable that the noise in many commercial photoelectric spectrophotometers is appreciably greater than the Johnson noise of the photocell load resistance. But while this will modify the numerical values calculated below, it in no way affects the nature of the argument.)

Suppose for a given wavelength and slit settings, the photoelectric current obtained with solvent in the beam is $i_0 = 10^{-10}$ ampere (this is a reasonable value for many commercial spectrophotometers).

At the 'optimum' absorbance of 0·43 the photoelectric current will be $i = 10^{-10} \cdot 10^{-0·43} = 10^{-10·43}$ ampere.

This gives [equation (4.15)]

$$\Delta A = 0·43 \cdot \Delta I/I \quad \text{(since again } i \propto I)$$
$$= 0·43 \cdot 10^{-15}/10^{-10·43} = 0·12 \cdot 10^{-4} = 0·000012$$

But this precision in A we know to be unrealistic and the error will be determined by conditions (b) above at not less than $\Delta A = 0·001$.

We shall in fact achieve a smaller fractional error in c by using a higher absorbance. For example, at $A = 2\cdot37$

$$\Delta i = 10^{-10} \cdot 10^{-2\cdot37}$$

which leads easily to

$$\Delta A = 0\cdot001$$

This absorbance is in fact the crossover point O of the two curves a and d of Fig. 4.7 and, in the conditions considered, is the optimum absorbance. The precision in concentration is then $0\cdot001/2\cdot37 = 0\cdot04$ per cent. For higher absorbance, the error will be greater, at point P say, on curve a, and for lower absorbance the error will also be greater, at, say, point Q on curve d.

But this optimum precision would not be obtained by comparing such a specimen with pure solvent, because for most instruments the absorbance scale is too cramped at this value. The specimen should be compared with a standard specimen of comparable concentration, reading at an open part of the absorbance scale; this is the method of difference spectrophotometry.

2. To take a second example, suppose that the wavelength and slit widths have to be chosen so that the current i_0 with solvent is only 10^{-13} ampere. Then similar calculations to those above lead to curve c of Fig. 4.7 and the optimum precision is attained at the lowest point of the upper of the two curves c and d, viz. at $A = 0\cdot43$. In this case, the optimum $\Delta c/c = \Delta A/A = $ [from equation (4.15)]: $(0\cdot43 \cdot 10^{-15})/(10^{-13} \cdot 0\cdot43 \cdot 10^{-0\cdot43}) = 10^{-1\cdot57} = 2\cdot7$ per cent. This may also be read off from the curve c.

Use of a photomultiplier

The advantage of a photomultiplier is that for a given primary photoelectric current the noise is much less than if the primary photo-current is amplified with a valve amplifier. In the latter case, the noise is Johnson noise arising in the high resistance in series with the photocell; but with a photomultiplier the noise is photon noise as discussed under heading 3 on page 37. As there shown [equation (4.11)] we may put

$$\Delta i \propto \sqrt{i} \tag{4.18}$$

i.e. the case is intermediate between those of equations (4.13) and (4.14).

Wybourne (1960) has shown by a calculation similar to that on page 75 that, if photon fluctuations are the only source of error, the optimum absorbance is twice that shown for (a) above; viz.

$$A_{\text{optimum}} = 0 \cdot 87 \tag{4.19}$$

The shape of the error curve is shown at e in Fig. 4.7, from which it is seen that the error increases more slowly with absorbance than is the case when Johnson noise applies.

2. PHOTOGRAPHIC PHOTOMETRY

Fig. 4.8 shows the characteristic curves of two types of film or plate. The ordinate—density of blackening b—is in fact the absorbance of the developed plate; and the abscissa is $\log I$ (sometimes $\log It$, where t is the exposure time). The gradient of the straight portion is called the plate gamma (γ) and its value is a measure of the contrast of the plate. Thus

$$\gamma = db/d(\log I) = db/0 \cdot 43(dI/I) \tag{4.20}$$

Photographic photometry consists in interpolating the blackening of an exposure between the blackenings of areas exposed to

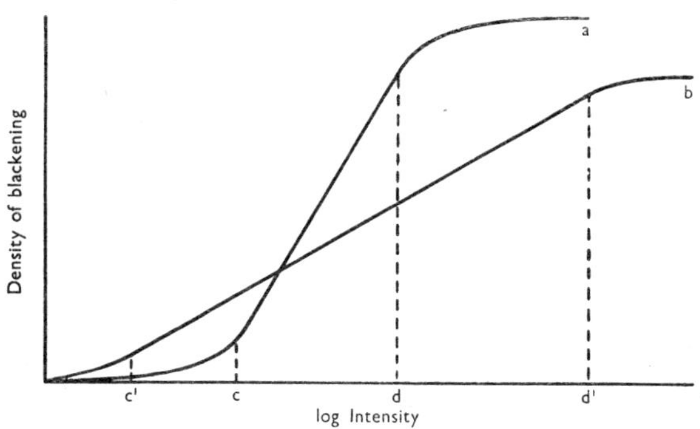

FIG. 4.8. Characteristic curves of two types of photographic emulsion.

known intensities; or, with a continuous range of known exposures, of locating areas where the plates are matched in blackening. An example of the latter method is shown in

Plate 3 (facing p. 48). The matching or interpolation may be done by visual inspection or better by using a photoelectric micro-photometer with the plate.

The exposure time may always be adjusted to give the optimum density of blackening b for such matching or interpolation, so that for a given plate we have $\Delta b =$ constant, i.e.

$$\Delta I/I = \text{constant}$$

This condition has already led us to the conclusion [equation (4.17) and curve d of Fig. 4.7] that a high absorbance of absorbing specimen gives the greatest accuracy.

Common-sense and equation (4.20) indicate that a plate with high contrast (large γ) will give a minimum $\Delta I/I$. Experience shows that with a high-contrast plate and a microphotometer, and using the precautions discussed below, the absorbance of an absorbing specimen may be measured to a precision of about $\Delta A = 0.003$ [see Follett (1935)].

Some comments on choice and use of plates or films

The Eberhard effect. During processing there is a region surrounding the dense portions in which development is retarded—the Eberhard effect ; this is caused by the local production of a large quantity of bromide which acts as a restrainer. Adequate agitation is therefore important but, if this is provided and development allowed to continue long enough to produce the greatest contrast (usually about four to five minutes), the Eberhard effect is usually unnoticeable.

Plate characteristics. In the wavelength range 2400–3200 Å the γ value of most plates remains constant at about 1 to 1·8, depending on the type of plate. At shorter wavelengths the γ value may be somewhat less, but at longer wavelengths it may increase to 9 or 10 for certain types of plate in the visible spectrum. A typical curve showing the relation of γ to wavelength is to be found in a paper by G. R. Harrison (1925). Plates of high γ (high contrast) are necessarily slow.

Developer. Ilford Ltd suggest caustic-hydroquinone developer (I.D. 13) as suitable for obtaining high contrast :

A	Hydroquinone	-	-	-	25 g
	Potassium metabisulphite		-		25 g
	Potassium bromide	-		-	25 g
	Water, up to -	-	-	-	1000 cm^3
B	Caustic potash	-	-	-	50 g
	Water, up to -	-	-	-	1000 cm^3

Use equal amounts of A and B.

3. VISUAL PHOTOMETRY

Nowadays, visual photometry merits only the briefest mention. The Weber-Fechner law tells us that the smallest change of intensity detectable by the eye is given by

$$\Delta I/I = \text{constant}$$

for intensities that are neither very low nor very high.

This condition again leads to the conclusion, as discussed in relation to equations (4.14) and (4.17) that a high value of absorbance gives $\Delta c/c$ of equation (4.1) its lowest value.

F

CHAPTER 5

Calculations : The Analysis of Mixtures

THE concentration of a specimen containing a single absorbing substance that is known to obey Beer's law may be determined merely by measuring its optical density at one suitable wavelength ; it can be assumed that the concentration is proportional to the measured absorbance. But analysis of mixtures of several absorbing substances is also possible in some cases even if Beer's law is not operative. This chapter sets out methods of calculating concentrations in various circumstances and shows what measurements have to be made. The following types of specimen will be considered:

1. A mixture of several substances in which the known absorption bands of each do not overlap.
2. A mixture of several substances in which the known absorptions overlap
   - (a) when Beer's law operates and
   - (b) when Beer's law does not operate.
3. A mixture in which some of the absorbing substances are unknown.

Methods will be illustrated by examples from published records.

Preparation for a routine analysis involves (i) choosing suitable wavelengths and (ii) calibrating by measurements of each of the pure components and tests on a few known mixtures. Procedure is simplified if one can choose a set of wavelengths with each showing large absorption for one component and zero absorption for the other components. This is not often possible and it is usually necessary to compromise with wavelengths showing the smallest possible absorption for other components. Choosing the most suitable wavelengths will reduce work and increase accuracy.

I. KNOWN ABSORPTION BANDS NOT OVERLAPPING

This gives the simplest conditions under which to analyse a mixture. If it is possible to choose a set of wavelengths at each of which one of the substances has sufficient absorption for accurate measurement and all the other substances have negligible absorption, then the absorption of each component can be measured

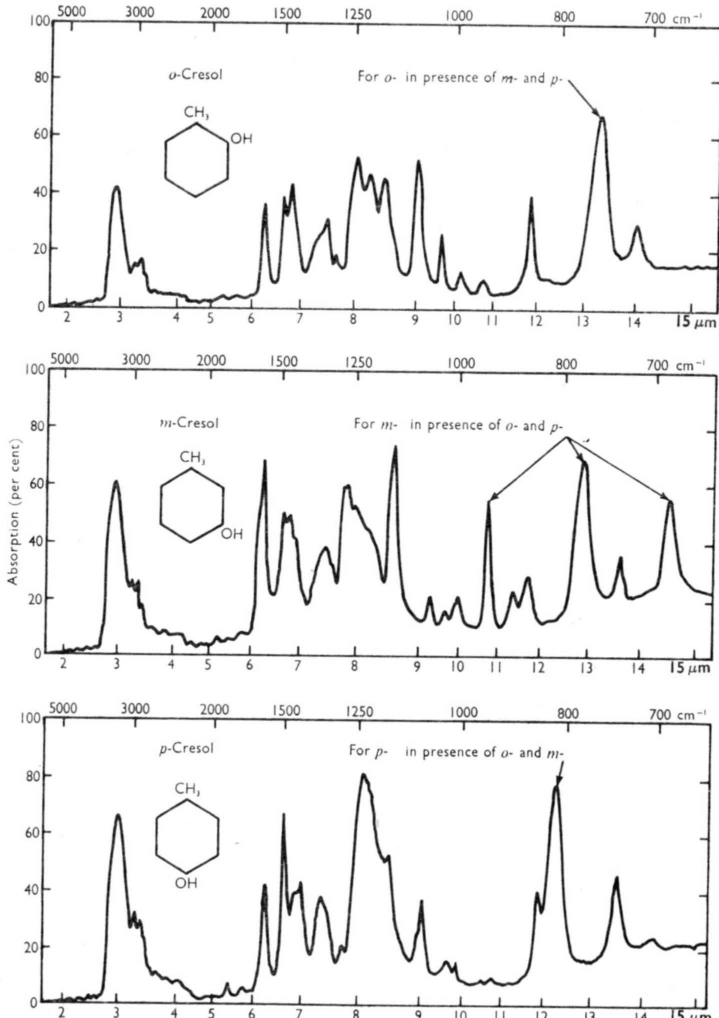

FIG. 5.1. Absorption curves of o-, m-, and p-cresol recorded on a Grubb-Parsons double-beam spectrophotometer using a rocksalt prism and specimens about 0·005 mm thick. (From Martin, 1952.)

independently of the others. (This does not apply where there is association or other form of chemical interaction as described on pages 23 and 122, or in making measurements on gases when there are pressure effects as described on page 33.)

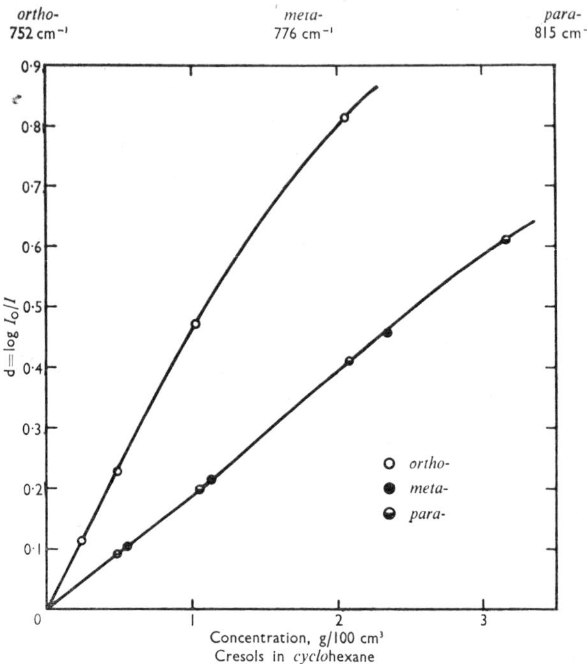

FIG. 5.2. Calibration curves for the estimation of cresols. (From Whiffen and Thompson, 1945.)

An example of this method is given by Whiffen and Thompson (1945). They describe the estimation of *ortho*-cresol, *meta*-cresol and *para*-cresol in mixtures of the three, an application which is finding some use in industry. It is possible to choose, for each of the three, a wavelength at which that particular component shows a peak of absorption while the other two show only negligible absorption. The wavebands used are in the 12–13 μm region (see Fig. 5.1). The applicability of Beer's law is tested by plotting curves of absorbance against concentration for each component separately—using solutions in carbon tetrachloride. The curves obtained are nearly straight, indicating only slight departure from

Beer's law. Nevertheless it is found preferable in this application not to use Beer's law in its algebraic form but to read off concentrations from the calibration curves, thus taking into account the small departure from linearity (see Fig. 5.2).

Results quoted on determinations made on synthetic mixtures showed agreement to about 2 per cent of the actual content. An example is :

	o-cresol	m-cresol	p-cresol
Percentage put into mixture	9·3	48·4	42·3
Percentage found	8·9	47·8	43·3

2. KNOWN ABSORPTION BANDS OVERLAPPING

(a) *When Beer's law operates*

If there are n components in the mixture the absorption of the mixture must be measured at n suitably chosen wavelengths ; the concentrations are then determined by solving n simultaneous equations. The algebraic method applicable to a three-component mixture is set out below.

Suppose that measurements are made at three wavelengths λ_1, λ_2, λ_3 on a mixture of three substances A, B and C ; that at the three wavelengths chosen the specific extinction coefficients are A : α_1, α_2, α_3, B : β_1, β_2, β_3 and C : γ_1, γ_2, γ_3 ; and that the extinction coefficients, for a mixture where the concentrations are c_A, c_B and c_C, are x_1, x_2 and x_3. Then, by an extension of Beer's law, the extinction coefficient of the mixture is the sum of the values for the separate components:

$$\left.\begin{aligned}
x_1 &= c_A\alpha_1 + c_B\beta_1 + c_C\gamma_1 \\
x_2 &= c_A\alpha_2 + c_B\beta_2 + c_C\gamma_2 \\
x_3 &= c_A\alpha_3 + c_B\beta_3 + c_C\gamma_3
\end{aligned}\right\} \qquad (5.1)$$

These three simultaneous equations may be solved for c_A, c_B and c_C. For greatest accuracy, wavelengths should be so chosen that at each the specific extinction coefficient of one component is much greater than that of the other two.

Nielsen and Smith (1943) have used equations of this type to estimate the composition of liquids containing three absorbing components during the manufacture of nitro-paraffins, the substances being various combinations of nitromethane, nitroethane,

1-nitropropane and 2-nitropropane. They were able to find infrared wavelengths at each of which one substance showed much larger absorption than the other three. This is indicated by the following figures quoted from their paper dealing with the specific extinction coefficients (for a concentration of one per cent by volume) of the separate substances.

Wave-length	Nitro-methane	Nitro-ethane	2-Nitro-propane	1-Nitro-propane
8·15 μm	0·252	0·486	0·177	**3·38**
10·06 μm	0·097	**1·85**	0·063	0·061
10·90 μm	**2·09**	0·114	0·135	0·782
11·74 μm	0·035	0·328	**5·26**	0·194

With these data they were able to make successful analyses.

(b) *When Beer's law does not operate*

Sometimes, when Beer's law is inoperative because the wavebands are too broad or because of other factors, and equations (5.1) cannot be applied, it is still possible to make use of spectrophotometric methods—if one can assume there is no interaction between the components. This problem has been treated in two different ways by Gordon and Powell (1945) and Fry, Nusbaum and Randall (1946). A summary of each method is given below.

The method used by Gordon and Powell requires an absorption cell of continuously adjustable length. In their paper, these workers show how such a cell may be used in the analysis of five-component hydrocarbon mixtures, and they set out a calculation in detail as an example ; in this book it is possible to give only an outline of the methods used and to summarize a typical calculation (see Table 5.1). Calculations are made not by solving simultaneous equations but by making successive approximations.

Column 1 of Table 5.1 shows the hydrocarbons to be analysed and column 2 shows the wavelengths chosen as giving maximum values of absorption for the corresponding hydrocarbon. (Benzene, for example, has large absorption at a wavelength of 9·64 μm but only small absorption at each of the other wavelengths mentioned—see Fig. 5.3.) Column 3 gives the absorbance for a thickness of 0·114 mm of the specimen under test.

Calibration graphs (as in Fig. 5.4) should have been previously constructed of absorbance against thickness for each component at each of the five wavelengths and the calculation by successive approximation is made somewhat as follows.

Assuming that at 13·80 μm all the absorption is due to n-hexane, the thickness of this solution is read from the appropriate calibration graph as 0·0813 mm. Absorbance of n-hexane at 9·64 μm due to this thickness is read from the appropriate curve as 0·070.

TABLE 5.1. Typical calculation for five-component hydrocarbon mixture

(1)	(2)	(3)	(4)	(5)	(6)	(7)	(8)
			First approximation		Second approximation		
Component	Wave-length (μm)	Absorb-ance measured	Thick-ness (mm)	Total absorb-ance at each wave-length	Thick-ness (mm)	Total absorb-ance at each wave-length	Volume (per cent)
Benzene	9·64	0·436	0·0178	0·486	0·0140	0·432	11·9
Methyl cyclo-pentane	10·24	0·190	0·0165	0·190	0·0178	0·185	15·2
3-Methyl pen-tane	13·00	0·150	0·0063	0·160	0·0063	0·142	5·4
2:2-Dimethyl pentane	13·50	0·252	0·0038	0·256	0·0102	0·258	8·6
n-Hexane	13·80	0·610	0·0813	0·672	0·0691	0·609	58·9

Subtracting from the observed absorbance at this wavelength, we get 0·436 − 0·070 = 0·366. From the calibration curves for benzene (Fig. 5.4) we can read, on the assumption that all this 0·366 is due to benzene absorption, that the thickness of benzene is 0·0178 mm. Proceeding in this way through all the wavelengths, we get the first approximations expressed as thicknesses in column 4.

The total absorbance at each wavelength is then calculated from these thicknesses and the values entered in column 5. Comparing these with the observed value of column 3 shows that the computed absorption at 13·80 μm is too high by 0·062, corresponding to a thickness of 0·0122 mm. This gives, as a second approximation, the thickness of n-hexane as 0·0813 − 0·0122

$=0.0691$ mm. A similar correction is made for benzene. The other three components are then recalculated for these new thicknesses of n-hexane and benzene.

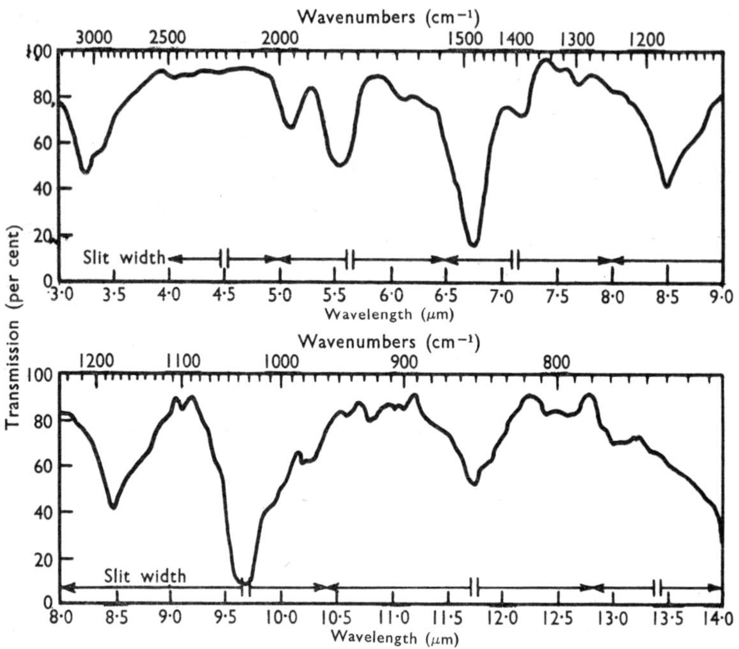

FIG. 5.3. Absorption of benzene in the infra-red.
(From Gordon and Powell, 1945.)

Column 6 shows the recalculated thicknesses and column 7 shows the total absorbance calculated for each wavelength; these values agree fairly well with the observed values of column 3. (The absorbances at $\lambda = 10.24$ μm and 13.00 μm are below 0.2 and, because such values cannot be measured accurately, check measurements and calculations are made on a greater cell thickness, when higher values are obtained. In this particular instance the check confirms the first observations.) If each thickness in column 6 is divided by the total calculated thickness we obtain the percentage composition by volume (column 8).

Fry, Nusbaum and Randall (1946) used a fixed-length cell

(length 0·152 mm) and gave examples of analysis of four-component and five-component mixtures. As usual, one wavelength for each component was required. Because they did not have an adjustable cell or a completely non-absorbing solvent,

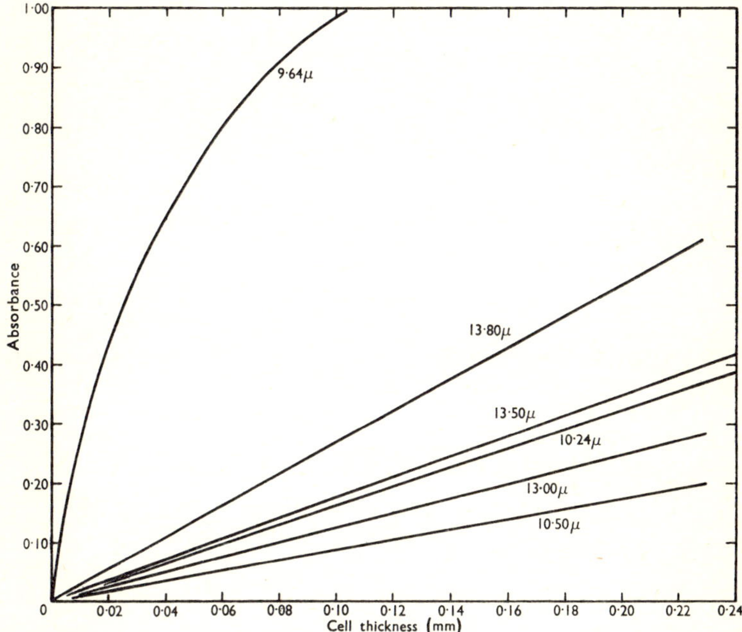

FIG. 5.4.   Calibration curves for benzene.
(From Gordon and Powell, 1945.)

these workers measured the absorbance of each component at various concentrations obtained by dissolving it in one of the other components showing only small absorption at the particular wavelength. The substances that made up the four-component mixture referred to in the paper are listed below, together with the corresponding wavelengths for maximum absorption.

Component	$\lambda(\mu m)$
2: 3-dimethylbutene-1 - - -	9·13
2: 2: 3-trimethylbutene - - -	11·89
Methylpentenes - - - -	12·30
2: 3-dimethylbutene-2 - -	10·31

Calibration curves were obtained by preparing solutions of various concentrations of each of the first three in 2 : 3-dimethylbutene-2. The latter was chosen because it has small absorption at all the wavelengths in question except for its peak at 10·31 μm.

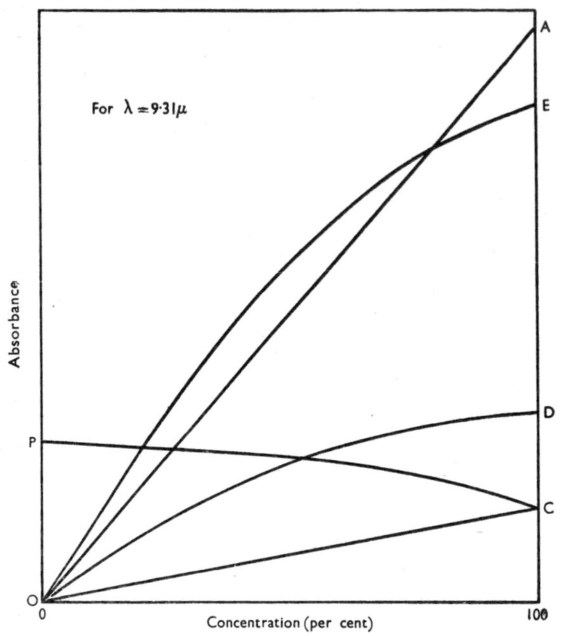

FIG. 5.5. Calibration curves
obtained by the method of Fry, Nusbaum and Randall (1946).

In Fig. 5.5, not to scale ($\lambda = 9\cdot31$ μm), the curve PC shows the optical density of various concentrations of 2 : 2 : 3-trimethyl-butene dissolved in 2 : 3-dimethylbutene-2, i.e. point C represents the optical density of pure 2 : 2 : 3-trimethylbutene at the working thickness of 0·152 mm, so that the line OC represents absorbance against concentration of this substance, independent of the absorption of any solvent. It is permissible to draw in OC as a straight line because, at the wavelength in question, we are not at an absorption peak of 2 : 2 : 3-trimethylbutene and departure from Beer's law will be small—also because any slight curvature of the line would have a negligible effect on the small absorbances in question.   (This substance is chosen for the straight line because it has the smallest absorbance at this wavelength.)

A curve OD, showing absorption of 2 : 3-dimethylbutene-2, can now be readily constructed from PC and OC and we can then construct graphs in the same way for each of the other pure substances at this wavelength, and likewise for other wavelengths. The method of successive approximation used by Fry, Nusbaum and Randall for computation from these calibration curves of an unknown mixture is very similar to that of Gordon and Powell just outlined. Fry, Nusbaum and Randall also give the following alternative method, which may often be convenient.

The first approximation is done algebraically, i.e. an approximate equation of the form $A = mc$ is written down for each of the calibration curves (sixteen for a four-component mixture) on the supposition that each is a straight line—OA, for example, is a straight line drawn through the curve OE.   Let A_1, A_2, A_3 and A_4 be the absorbances measured on the unknown mixture at the four wavelengths λ_1, λ_2, λ_3 and λ_4, let c_A, c_B, c_C and c_D be the four concentrations to be determined, and let m_1 to m_{16} be the gradients of the approximate straight lines.   Then

$$\left.\begin{aligned}
A_1 &= m_1 c_A + m_2 c_B + m_3 c_C + m_4 c_D \\
A_2 &= m_5 c_A + m_6 c_B + m_7 c_C + m_8 c_D \\
A_3 &= m_9 c_A + m_{10} c_B + m_{11} c_C + m_{12} c_D \\
A_4 &= m_{13} c_A + m_{14} c_B + m_{15} c_C + m_{16} c_D
\end{aligned}\right\} \qquad (5.2)$$

For routine use, solutions of these equations may be written down and put into the simplest numerical form for calculation, e.g.

$$\left.\begin{aligned}
c_A &= \alpha_1 A_1 + \alpha_2 A_2 + \alpha_3 A_3 + \alpha_4 A_4 \\
c_B &= \alpha_5 A_1 + \alpha_6 A_2 + \alpha_7 A_3 + \alpha_8 A_4 \\
c_C &= \alpha_9 A_1 + \alpha_{10} A_2 + \alpha_{11} A_3 + \alpha_{12} A_4 \\
c_D &= \alpha_{13} A_1 + \alpha_{14} A_2 + \alpha_{15} A_3 + \alpha_{16} A_4
\end{aligned}\right\} \qquad (5.3)$$

First approximation values of concentrations may be calculated by using equations (5.3).*

The second approximation is made by using a set of curves that

---

* Transformation from equations (5.2) to (5.3) is convenient if the determinant method of solution is used ; this is described in most textbooks on higher algebra and is well set out in Turnbull's *Matrices, Determinants and Invariants* (Blackie, 1948).

show concentrations as abscissae; the ordinates are the absorbance differences between the actual calibration curves and the straight lines drawn through them for the first approximation. Such correction curves for the four components at one wavelength may have the form indicated in Fig. 5.6. The curves may be constructed with an open absorbance scale without having an inconveniently large sheet, and this seems to be one of the great advantages of this method.

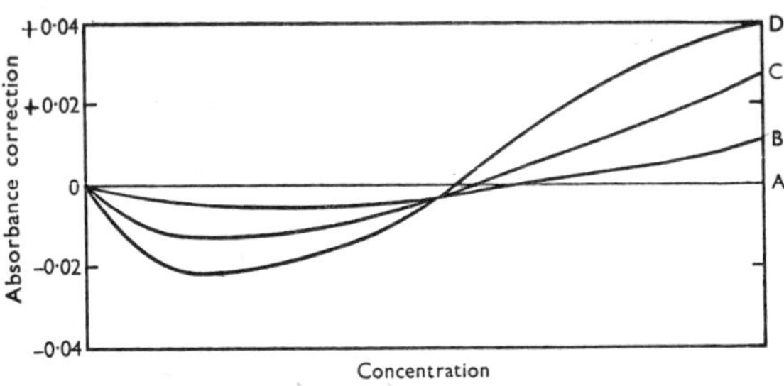

FIG. 5.6. Correction curves for the method of Fry, Nusbaum and Randall (1946).

Using the four concentrations determined at one wavelength (say λ_1) in the first approximation, four corrections are read from the correction curves of Fig. 5.6. The corrections are added (taking account of sign) and the resulting absorbance correction added to or subtracted from the measured value A_1 to obtain a corrected A_1. This process is then applied to each of the other wavelengths to obtain corrected values A_2, A_3 and A_4. The new values are substituted into equations (5.3) and second approximations calculated for concentrations.

The whole process must now be repeated (third approximation, etc.) until the last approximation gives the same set of concentrations as the previous one. Experience will show how many approximations are required for a given set of calibration curves. The less the curvature of these, i.e. the smaller the departure from Beer's law, the fewer the approximations required. For the four-component mixture mentioned above, three approximations were found sufficient.

3. ELIMINATING THE EFFECT OF UNKNOWN ABSORPTION BANDS

It is exceptional for biological specimens, in particular, to contain only known substances. For example, in estimating substances such as vitamin A, vitamin B, vitamin D, etc., it usually happens that there are other unknown substances present whose absorptions overlap that of the vitamin; in such cases a spectrophotometric measurement indicates too large a concentration. This problem may be dealt with in one of a number of ways. Two chemical methods will be mentioned in terms of the estimation of vitamin A, other methods will be discussed in a more general manner.

1. The impurities may be removed by chemical processes. In fish-liver oils and butter the fats can be removed by saponification and a spectrophotometric measurement made on the remaining non-saponifiable portion. A more accurate measurement, still tending to be high, will then be obtained.

2. The vitamin A may be destroyed in such a way as to leave the remaining substances unchanged. Under certain conditions the vitamin A will be destroyed if it is irradiated; the new (decreased) absorption can then be measured and subtracted to correct the original measurement.

But such chemical manipulation can sometimes be avoided by suitable operations on an observed spectrophotometric curve. The principle of all such methods is that the ' shape ' of an absorption band is characteristic of a given pure substance. The shape of a band is characterized by the ratios of pairs of absorbances at various pairs of wavelengths. If in an observed curve the values of such ratios are different from those for a pure substance, the presence of impurities is indicated. It is sometimes possible to make use of this change of band shape to determine the concentration of the pure substance. There are several alternative possibilities:

(a) Linear impurity absorption

Correction for the effect of impurities is fairly simple when their absorbance varies linearly with wavelength. Probably the first person to report the use of such a method was Wright (1941), who showed how to subtract the background absorption at the band peak of a substance to be estimated by extending the smooth curve from either side of the band. This is sometimes known as the base-line method. This principle has since been elaborated in various ways to eliminate the need for actual curve construction. Among the earliest of such methods were those of Banes and Eby (1946) and Morton and Stubbs (1946). The various methods have been reviewed by Mulder, Spruit and Keuning (1963) and White (1964). The use of one such method is described in Chapter 7, p. 140 (estimation of paraquat and diquat). The treatment of Morton and Stubbs, which they developed in connection with the estimation of vitamin A, will now be discussed more fully.

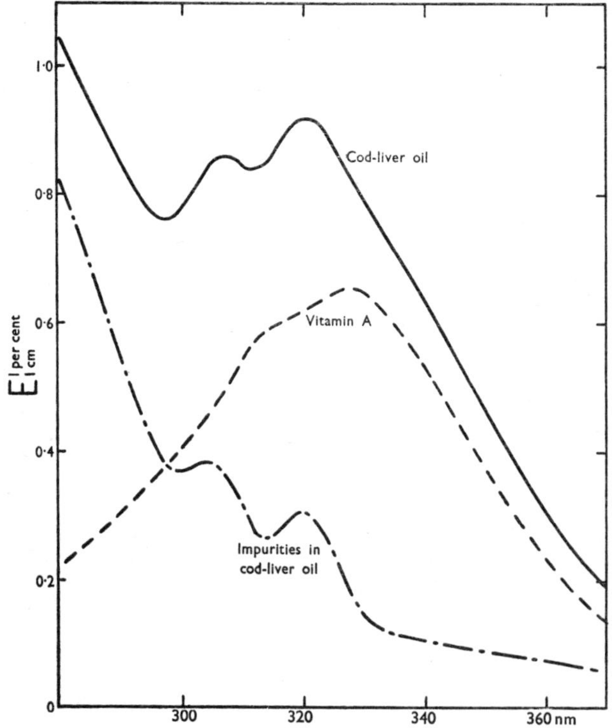

FIG. 5.7. The absorption of vitamin A, cod-liver oil, and impurities in cod-liver oil. (From Morton and Stubbs, 1946.)

Morton and Stubbs method. In Fig. 5.7, taken fron their paper, the broken curve shows the absorption of relatively pure vitamin A, the unsaponifiable fraction of cod liver oil; the full curve, however, shows the absorption measured for a specimen of cod liver oil, and the changed shape obviously indicates the presence of other absorbing substances. Provided the impurities do not have peaks in the region of 3280 Å, it is possible, making only the reasonable assumption that the impurity absorption is linear over a small wavelength range in this region, to calculate the concentration of the pure substance.

The simplest form of the calculation may be followed from Fig. 5.8. The upper dotted curve represents the absorption of the pure substance and the full-line curve represents the measurements of the absorption of the substance plus irrelevant absorption, the

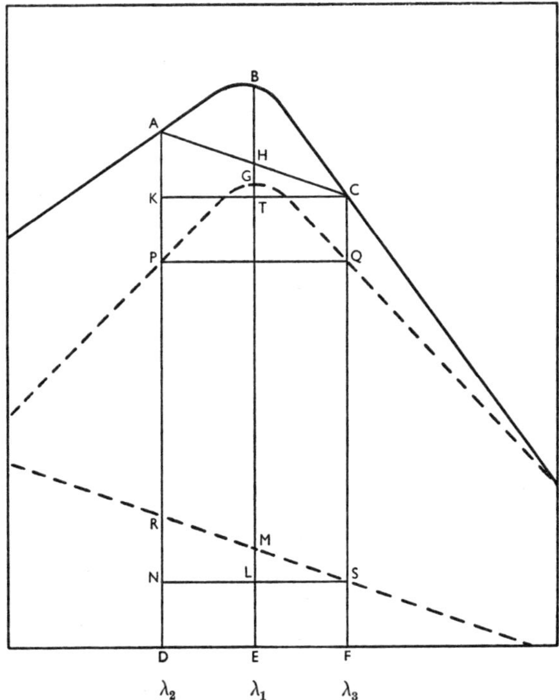

FIG. 5.8.   Illustration of the method of calculation
used by Morton and Stubbs (1946).

lowest curve RMS representing the difference between the two namely the irrelevant absorption.

Three wavelengths are chosen at D, E and F. The point E is at the absorption peak of the pure substance and D and F are at wavelengths on either side of the peak at which the extinction coefficients of the pure substance are equal. If the ratio of the extinction coefficients of the pure substance at wavelengths E and F be denoted by r, then it follows from the geometry of Fig. 5.8 that

$$r = \frac{GE}{QF} = \frac{BE - BG}{CF - CQ}$$

and therefore, since $BG = ME = ML + LE = HT + CQ$,

$$r = \frac{BE - HT - CQ}{CF - CQ} \tag{5.4}$$

The value of r is known for the pure substance, BE and CF are the measured absorbances, and HT is equal (see figure) to half the difference of the measured absorbances at wavelengths D and F; so that CQ is the only unknown in this expression and its value can be calculated. This gives a value for QF, the absorbance of the pure substance, from which its concentration can be calculated.

Morton and Stubbs have applied this procedure to vitamin A in liver oils, using 3130, 3280 and 3385 Å for the three wavelengths. It will be seen from Fig. 5.7 that the irrelevant absorption shows some peaks (attributed to conjugated tetra-ene acids) and it is therefore important to choose carefully the three wavelengths to give linearity in the irrelevant absorption.

A more useful form for the relationship between the corrected value for the peak absorbance ($A_{1,\text{corrected}}$) and the observed quantities is easily deduced from expression (5.4), or in some alternative manner:

$$A_{1,\text{corr}} = \left(\frac{r}{r-1}\right)\left[A_1 - A_3 - (A_2 - A_3)\frac{(\lambda_3 - \lambda_1)}{(\lambda_3 - \lambda_2)}\right] \tag{5.5}$$

where $r = A_1/A_3 = A_1/A_2$ for the pure substance.

Shaw and Jefferies (1953) have given an alternative expression for the more general case when the absorbances of the pure substances at the subsidiary wavelengths λ_2, λ_3 are not equal, but for which

$$r_2 = A_1/A_2, \qquad r_3 = A_1/A_3$$

Their expression is:

$$A_{1,\text{corr}} = \frac{A_1 r_2 r_3(\lambda_3 - \lambda_2) - A_2 r_2 r_3(\lambda_3 - \lambda_1) - A_3 r_2 r_3(\lambda_1 - \lambda_2)}{r_2(\lambda_2 - \lambda_1) + r_2 r_3(\lambda_3 - \lambda_2) + r_3(\lambda_1 - \lambda_3)} \qquad (5.6)$$

Shaw and Jefferies used the Morton and Stubbs method for the estimation of ergosterol by means of its absorption band at $\lambda = 2820$ Å. This band is however fairly narrow with steep sides, and a wavelength setting error of only 5 Å could give a concentration error of 19 per cent. They found it better to use for the subsidiary readings values of absorbance at adjacent maxima; their paper may be consulted for details.

(b) *Non-linear impurity absorption.*

The absorption of impurities may not vary linearly with wavelength and more general cases need terms in λ^2, λ^3, etc., in the A_λ/impurity-concentration relation. Ashton and Tootill (1956) described a method of dealing with such impurities. It is fairly obvious that measurements must be made at a minimum of four wavelengths; but in applying it to the estimation of griseofulvin by means of its absorption peak at $\lambda = 2892{\cdot}5$ Å, the above authors found it better to use seven wavelengths—2880, 2900, 2920, 2940, 2960, 2980 and 3000 Å. Their paper should be consulted for details of the calculation, which involves the use of orthogonal polynomials taken from page 90 of the statistical tables by Fisher and Yates (1957).

Glenn and collaborators regard absorbing impurities as equivalent to 'noise'. To improve the signal-to-noise ratio, they narrow the pass band by basing the concentration of pure compound on just one orthogonal function coefficient in place of the combination of such coefficients used by Ashton and Toothill. This approach leads to simple arithmetic and minimum assumptions about the impurity absorption curve.

Any experimental curve of absorbance v. wavelength can be expanded in the form

$$A(\lambda) = p_0 P_0(\lambda) + p_1 P_1(\lambda) + p_2 P_2(\lambda) + \ldots + p_n P_n(\lambda)$$

G

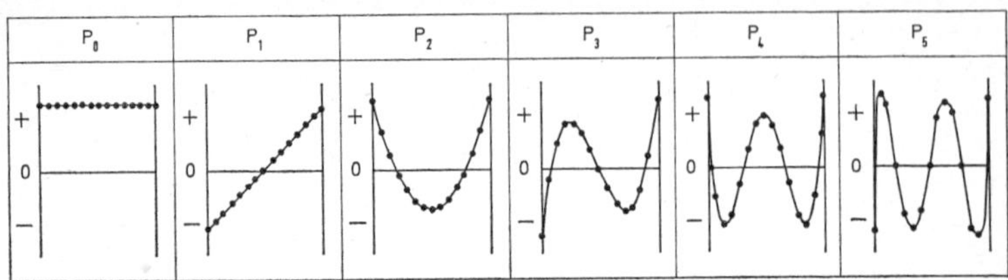

FIG. 5.9. Orthogonal polynomials. In each diagram, $P(\lambda)$ is plotted against a set of equally spaced wavelengths, the same wavelengths being used for all six diagrams. Each set of points represents a fundamental curve shape, whose general characteristics are indicated by the line joining them.

where [to quote from Glenn (1967)] $A(\lambda)$ is one of

'a set of optical density values [i.e. absorbances] at $(n+1)$ equally spaced wavelengths at which the orthogonal polynomials $P_0(\lambda)$, $P_1(\lambda)$, $P_2(\lambda)$ etc. are each defined. These polynomials represent a series of fundamental curve shapes [see Fig. 5.9] that may be scaled up or down by the appropriate coefficients (p_0, p_1, p_2, etc.) and then summed to reproduce the value of any $A(\lambda)$ within the set. In applying this method to the determination of an absorbing compound in the presence of irrelevant absorption, each coefficient, like $A(\lambda)$ itself, represents the sum of contributions from background and absorbing compound, respectively. However the ratio of the two contributions varies from one coefficient to another, and by suitable choice of wavelengths it is usually possible to find one particular coefficient (e.g., p_2) to which the background contributes a negligible fraction of the whole. That coefficient is then proportional to the concentration of the absorbing compound. Further, as any background must contribute to p_0 and probably to p_1 also, the analytical result is always based upon p_2 or some higher coefficient.'

This method is discussed by Glenn (1963) in relation to the determination of two component mixtures and by Agwu and Glenn (1967) as a method of estimating purity. Glenn and Wahbi (1969) discuss the method for the estimation of a single substance in the presence of impurities and stress that the success of all these methods depends on the careful and correct choice of conditions.

PART II

Applications of Spectrophotometry

CHAPTER 6

The Investigation of Chemical Structure and Chemical Reaction

---

THIS chapter will discuss in an elementary way the relationship between absorption spectra and chemical structure, and will deal with both electronic (ultra-violet) and vibration-rotation (infra-red) absorption. A few examples will be given of the ways in which absorption spectrophotometry may be used in elucidating the structure of compounds and in considering molecular moments of inertia, photochemical reactions, and chemical equilibria. The treatment of this large and growing subject will necessarily be very brief and the reader is referred to the bibliography for sources of more detailed information.

ULTRA-VIOLET SPECTRA

THE STRUCTURE OF COMPOUNDS

Chromophoric groups

As explained in Chapter 1, Hartley (1883), Victor Henri (1919) and Baly (1927) noticed examples of systematic relationships between the absorption spectra of a compound and its chemical structure. It was shown that the absorption bands of a compound can be attributed to certain groupings of the molecule known as ' chromophoric ' groups. The $C=O$ group characteristic of

TABLE 6.1. Absorption bands of some chromophoric groups

Group	λ max. (Å)	ϵ	λ max. (Å)	ϵ
$C=O$*	1870	1000	2800–3200	17
$C=C$	1950	10,000	2300–2500†	
$O-H$			2300	0·2
$C-C$			1350	
$C-H$			1250	

* Aldehyde or ketone.
† Not a true maximum but a kink or ' step-out '.

aldehydes and ketones, for example, gives rise to an absorption that is usually in the region of 2800–3200 Å (see Table 6.2).

Other chromophoric groups have their own characteristic absorption bands as indicated in Table 6.1. The approximate intensities given are expressed as values of molecular extinction coefficient (ϵ).

The absorption maximum of a chromophoric group in a particular molecule is influenced by the rest of the molecule. Thus the following absorption maxima of several compounds containing the $C{=}O$ group are all attributed to a transition from a non-bonding orbital on the oxygen atom to an anti-bonding orbital in the $C{=}O$ bond.

TABLE 6.2. Absorption bands of compounds with $C{=}O$ groups

Carbon dioxide	CO_2	1950 Å
Acetic acid	CH_3COOH	1950 Å
Esters	$RCOOR$	$\sim$2100 Å
Acetyl chloride	CH_3COCl	$\sim$2800 Å
Formaldehyde	$HCHO$	$\sim$2900 Å
Acetaldehyde	CH_3CHO	2900 Å
Acetone	$(CH_3)_2CO$	$\sim$3000 Å
Benzaldehyde	C_6H_5CHO	3200 Å

A band in the 2800–3200 Å range is typical of aldehydes and ketones, but in other molecular forms the band arising from the same electronic change moves considerably. Walsh (1947) has shown that there is a correlation between the wavelength and the polarity of the bond as determined by ionization potentials or dipole moments ; longer wavelengths are associated with greater polarity.

Where a molecule includes several chromophoric groups not too close together (unconjugated, as defined below) the resulting absorption spectrum is given approximately—both in position and intensity—by addition of the bands characterizing the individual groups. Thus acetonyl acetone, $CH_3COCH_2CH_2COCH_3$, with two $C{=}O$ groups separated by two carbon atoms, has a relatively weak band at 2700 Å ($\epsilon = 120$) in about the same position as acetone and more than twice as intense (Henri, 1919). Similarly methyl heptenone, $(CH_3)_2C{=}CH \cdot CH_2CH_2COCH_3$, with $C{=}C$ and $C{=}O$ groups separated by two carbon atoms, has bands near 1950 and 2750 Å characteristic of the separate chromophoric groups.

CONJUGATED GROUPS

When two groups are closer together in the molecule, separated by only one carbon atom, they are said to be conjugated. As long ago as 1889, Thiele recognized that such molecules have special chemical properties. When two or more groups are conjugated, the absorption bands cease to be those of the separate groups and new bands appear at longer wavelengths. For example, for mesityl oxide, $(CH_3)_2C{=}CH{-}C{=}O . CH_3$, in which the $C{=}C$ and $C{=}O$ chromophoric groups are conjugated, we no longer have the bands corresponding to the separate groups at 1950 and 2750 Å, but new bands appear at longer wavelengths, namely 2300 and 3270 Å.

When more than two groups are conjugated there is a shift to successively longer wavelengths as the number of groups increases. Hausser, Kuhn and others (1935), in an important set of papers, report on the absorption spectra of several series of compounds. Table 6.3, which gives absorption peaks compiled from these papers and other sources, shows (a) the gradual increase of wavelength with increasing number of double bonds (reading down) and (b) the effect of the rest of the molecule (reading across); the addition of an OH group (column 2) causes a shift to shorter wavelengths. Although, in the third column of the table, two of the $C{=}C$ bonds form part of a ring system, the same general rules apply.

TABLE 6.3. Absorption maxima of compounds with conjugated $C{=}C$ groups

$CH_3{-}(CH{=}CH)n{-}CHO$	$CH_3{-}(CH{=}CH)n{=}COOH$	⬡O$-(CH{=}CH)n_{-2}{-}CHO$
$n=1$ 2170 Å	2040 Å	—
2 2700 Å	2540 Å	2720 Å
3 3160 Å	2940 Å	3140 Å
4 —	3270 Å	3500 Å
5 —	—	3730 Å

We can here outline the properties of only these one or two important chromophoric groups. More extensive information may be obtained from Morton (1942), who summarizes the subject with a number of tables and graphs showing absorption maxima for several groups of compounds. Fairly extensive tables may also be found in a paper by Dimroth (1939).

Wave-mechanical theory of conjugation

Lowry and Allsopp (1937) found that the shift in absorption bands when a pair of double bonds are separated by only one single bond depends on whether the bonds are in a *cis* or *trans* position, and that the mere conjugation or closeness of bonds is not sufficient to produce the shift. Following these observations, Mulliken (1939) published an important series of papers setting out the wave-mechanical ideas of resonance between such bonds. A full theoretical treatment is not possible in this book, but a brief outline of current ideas may help to give a balanced picture of the subject.

In the double bond $C=C$, one pair of the four electrons forming the covalent link is regarded as mobile and moving in the field of the two carbon atoms. In conjugated bonds, each mobile electron moves through the field of the whole nuclear system and of the other mobile electrons (or, in the older phraseology, the mobile electrons move in orbits which enclose all the carbon atoms of the conjugated groups). Such electrons are less strongly bound than those rotating around single atoms and the energy levels are thus closer together, i.e. the absorption wavelengths are greater for conjugated bonds than for single bonds. These ideas are discussed in various textbooks on organic chemistry—e.g. Gillam and Stern (1960).

APPLICATIONS TO STRUCTURAL PROBLEMS

We will now consider a few typical examples of how the relationship between the ultra-violet absorption and chemical structure may be used in spectrophotometry. It is necessary, however, to be very cautious about assuming structural relationships from observed absorption bands; the subject is still growing and our knowledge of it is incomplete. The absorption spectra of iodine compounds, for example, show how easy it would be to draw wrong conclusions. Both CHI_3 and KI_3

have bands at 2900 and 3500 Å and this suggests that these compounds have a common type of bonding. Allsopp (1937), however, compared the bands for various other iodine compounds, including AsI_3 and $SnCH_3I_3$, as well as I_2 adsorbed on a calcium fluoride surface. He found that, in common with CHI_3 and KI_3, all possessed two bands with a separation of about 0·8 electron-volts, although chemical considerations showed that the structures could not all be similar. After a series of careful experiments, Allsopp concluded that the common frequency difference must be attributed to simultaneous production by the incident radiation of iodine atoms in the normal and the excited metastable states—one band corresponding to the change $XI_2 \rightarrow X + I + I$ and the other to the change $XI_2 \rightarrow X + I + I^\star$.

Structure of polyphenyl compounds

Gillam and Hey (1939) have measured the absorption spectra of polyphenyl compounds in the *para* and *meta* series and have reached some important conclusions on the nature of the bonding. Their results, showing λ_{max} and the corresponding values of molar absorptivity in chloroform solution, are quoted in Table 6.4:

TABLE 6.4. Absorption data for polyphenyl compounds

Compound		λ_{max}	ϵ_{max}
Para series			
Diphenyl	$C_{12}H_{10}$	2515 Å	18,300
Terphenyl	$C_{18}H_{14}$	2800	25,000
Quaterphenyl	$C_{24}H_{18}$	3000	39,000
Quinquiphenyl	$C_{30}H_{22}$	3100	62,000
Sexiphenyl	$C_{36}H_{26}$	3175	$>56,000$
Meta series			
Diphenyl	$C_{12}H_{10}$	2515	18,300
Terphenyl	$C_{18}H_{14}$	2515	44,000
Noniphenyl	$C_{54}H_{38}$	2530	184,000
Deciphenyl	$C_{60}H_{42}$	2530	213,000
Undeciphenyl	$C_{66}H_{46}$	2530	215,000
Duodeciphenyl	$C_{72}H_{50}$	2530	233,000
Tredeciphenyl	$C_{78}H_{54}$	2530	252,000
Quatuordeciphenyl	$C_{84}H_{58}$	2530	283,000
Quindeciphenyl	$C_{90}H_{62}$	2540	309,000
Sedeciphenyl	$C_{96}H_{66}$	2550	$\sim$320,000

In the *para* series, the increasing wavelengths of the absorption band, together with the increasing intensities as the number of benzene rings increases, are an indication of a series of conjugated double bonds, the structure being

The fact that X-ray analysis shows that the molecules lie in a plane also suggests that the structure is of this type.

In the *meta* series on the other hand, because of the valency requirements of the carbon atoms, simple conjugation in the conventional sense cannot extend beyond that inherent in the diphenyl structure ; the compounds of the *meta* series of Table 6.4 that contain n benzene nuclei should behave as $n/2$ independent conjugated chromophores. We should therefore expect the absorption maximum to remain near the wavelength corresponding to diphenyl (i.e. 2515 Å) and the intensity to increase uniformly as further nuclei are added. The table shows that this is so. If the effective chromophore were diphenyl, one would expect a uniform increase of intensity for the addition of each further pair of benzene nuclei, whereas the table shows that there is a regular increase of intensity when each single nucleus is added. Gillam and Hey therefore conclude that the effective chromophore is $\frac{1}{2}(C_6H_5)_2$.

APPLICATIONS TO PHOTOCHEMISTRY

There are several ways in which absorption spectrophotometry can be of use in the study of photochemical reactions initiated when radiation is absorbed. Heilbron, Kamm and Morton (1927) found, for instance, that irradiation of impure cholesterol gives rise to a photochemical change producing a substance with antirachitic properties. Pure cholesterol, however, is transparent in the 2000–3000 Å region and so could not be the source of the antirachitic substance. Fractional crystallization separated out from the impure cholesterol one substance with absorption bands at 2935, 2815, 2700 and 2600 Å. This component was obviously the source of the photochemical change and was found to be a progenitor of vitamin D. The absorption measurements in this instance also served to show that there was no point in irradiating with wavelengths outside the absorbing region (2500–3000 Å) of the provitamin.

Photochemical decomposition of iodine (I_2)

The absorption spectrum of iodine vapour shows a band system in the visible spectrum (Plate 1), but at $\lambda \leqslant 4995$ Å the absorption becomes continuous. At this wavelength the absorption corresponds to a transition indicated by C in the Franck-Condon diagram (Fig. 2.2b), i.e., the quantum of energy at this wavelength is just equal to the dissociation energy. Using the conversion factor given in Table A, it is seen that the energy associated with N_A photons of $\lambda = 4995$ Å is $2 \cdot 85 . \ 10^8/4995 = 56,950$ calories per gram-molecule. It can be deduced from spectroscopic evidence that the energy required to excite an iodine atom is 21,650 calories per gram atom. If the iodine molecule is dissociated into one normal and one excited atom, the heat of dissociation should be $56,950 - 21,650 = 35,300$ calories per gram-molecule. The heat of dissociation obtained thermochemically is 35,100 calories, so that the explanation given above appears to be valid. It is also supported by other spectroscopic evidence.

Photochemical efficiency

The decomposition of gaseous hydrogen iodide to hydrogen and iodine by ultra-violet radiation gives a further example. The spectrum of HI shows continuous absorption below about 3000 Å and the initial stage of the reaction is regarded as

$$HI \rightarrow H + I$$

Measurements of absorbed radiation, and of the amount of hydrogen and iodine produced, show that for every quantum of ultra-violet radiation absorbed, approximately two molecules of HI are decomposed; in other words the quantum efficiency is 2. This must obviously be ascribed to further chemical reactions, and detailed consideration leads to the reactions

$$HI + h\nu \rightarrow H + I$$
$$H + HI \rightarrow H_2 + I$$
$$I + I \rightarrow I_2$$

The kinetics of a photolytic reaction

Flash photolysis of a solution of iodine in benzene by radiation in the visible spectrum gives rise to the decomposition

$$I_2 \rightarrow I + I$$

followed by the formation of a complex C_6H_6I; this latter has an absorption in the violet region. The complex decays in a few milliseconds and the rate of decay may be followed [Rand and Strong (1960)] by observation at $\lambda = 4200$ Å where the absorptions of I_2 and a stable complex $C_6H_6I_2$ are small. A separate beam from a tungsten-filament lamp is used with an interference filter, photomultiplier and cathode-ray oscilloscope. The shape of the decay curve indicates a second-order reaction, i.e.,

$$2C_6H_6I \rightarrow C_6H_6I_2 + C_6H_6$$

with the rate of decay k determined by

$$(d/d)[C_6H_6I] = -2k[C_6H_6I]^2$$

Dissociation of the oxygen molecule

Oxygen absorbs at wavelengths a little shorter than 2000 Å and spectrometers for such wavelengths therefore have to be evacuated. Fig. 6.1 shows the Schumann bands of oxygen (O_2). This band system arises from a single electronic transition associated with various vibrational quantum jumps (numbered from 8 to 20 in the figure) with still finer structure due to rota-

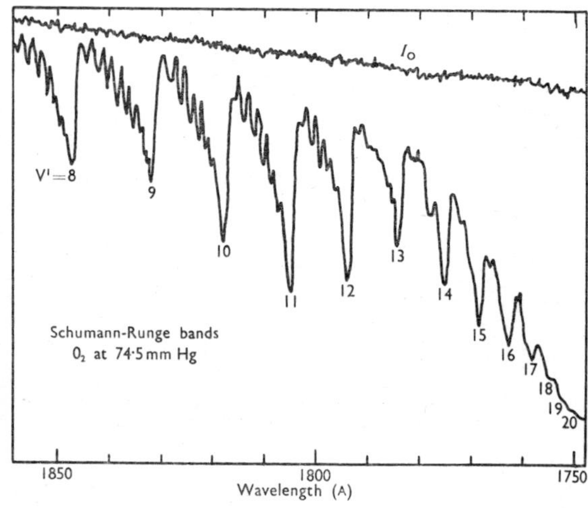

FIG. 6.1. The Schumann-Runge absorption bands of oxygen recorded on a photoelectric grating spectrophotometer described on page 182. (From Inn, 1955.)

tional transitions. It is seen that the vibrational structure is converging towards a limit near 1750 Å; by careful extrapolation Brix and Herzberg (1954) find the exact convergence limit corresponding to dissociation of the molecule to be at 1750·5 Å or 57,128 cm^{-1}. This dissociation leaves the oxygen atoms in an excited state with an energy of 15,868 cm^{-1} above the ground state. Consequently the energy required to dissociate the O_2 molecule into two unexcited atoms is $57,128 - 15,868 = 41,260$ cm^{-1}, which is equivalent to 5·115 electron volts or 117,960 calories per gram-molecule. Fig. 6.1 shows that the bands of highest intensity are those for high values of v', the vibrational quantum number of the upper state. This indicates that the Franck-Condon curves of Fig. 2.2b apply rather than those of Fig. 2.2a, i.e., the bond length is appreciably increased in the upper electronic state.

MISCELLANEOUS APPLICATIONS

Variation of Molecular Dimensions Deduced from Fluorescence Spectra

Referring again to page 15 and to the Franck-Condon diagrams of Fig. 2.2, it was there explained that, if interatomic distances in a molecule are the same in ground and electronically excited states (Fig. 2.2a), the most probable absorption transition is $v''=0 \rightarrow v'=0$; and in that case the strongest fluorescent emission is also $v'=0 \rightarrow v''=0$. That is, the strongest absorption band and the strongest fluorescence band will coincide; this is seen to be the case for anthracene (Fig. 6.2b). But where Fig. 2.2b applies, the most probable absorption transition lies at a shorter wavelength than the most probable fluorescence emission. Obviously this latter state of affairs applies to the molecules fluorescein, diphenyloctatetraene and benzene illustrated in Fig. 6.2a, c and d. Thus in these last molecules the interatomic distances are obviously greater in the upper electronic states concerned.

Chemical equilibria

A method developed by Theorell and Chance (1951) illustrates some useful techniques. These workers used absorption measurements to investigate the reversible combination of an enzyme, alcohol-dehydrogenase (ADH) from horse liver, with reduced diphosphopyridine nucleotide (DPNH), i.e.

$$\text{ADH} + \text{DPNH} \rightleftharpoons \text{ADH—DPNH}.$$

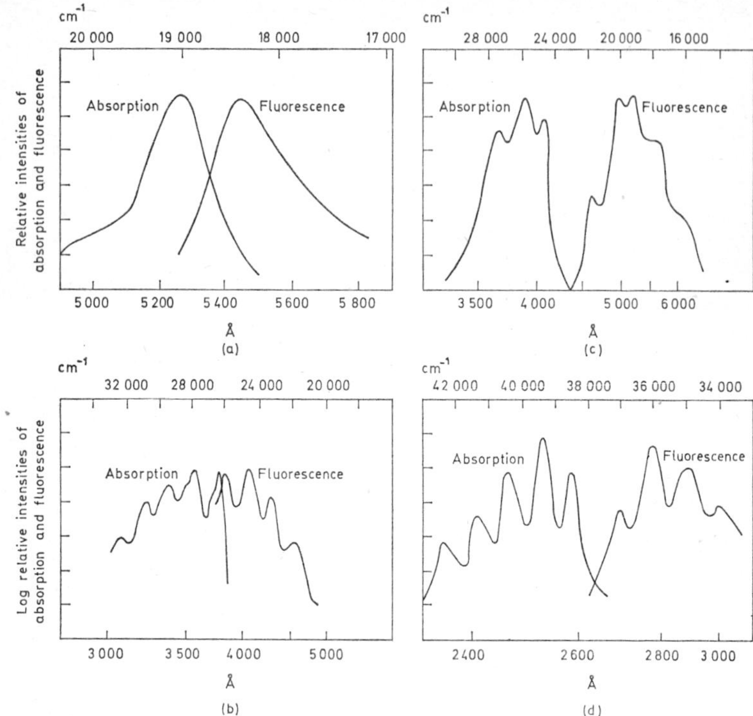

FIG. 6.2. Approximate mirror image relation between absorption and fluorescence spectra. (*a*) Fluorescein; (*b*) Anthracene; (*c*) Diphenylocta-tetraene; (*d*) Benzene. (*a, b, c* are reproduced from West, W. (1956).)

The absorption curves (Fig. 6.3) show that the absorption of free and ADH-bound DPNH is constant at 328 nm, i.e. $\lambda = 328$ nm is an isosbestic point for the two compounds. Free DPNH has the same absorption at 328 and 354 nm and therefore

(*a*) the total DPNH is determined by d_{328}, the absorbance at 328 nm, and

(*b*) the ADH-bound DPNH is measured by $d_{328} - d_{354}$.

The reaction mixture is irradiated with the two wavelengths alternately, by means of two monochromators and a vibrating mirror. Using a photocell and suitable circuits, and switches

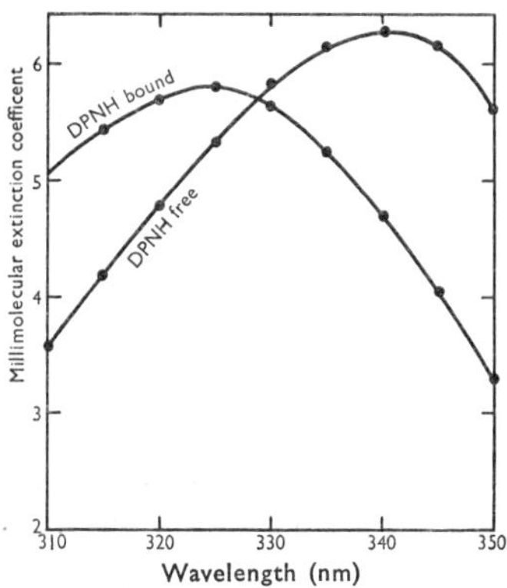

FIG. 6.3. Absorption curves from Theorell and Bonnichsen (1951).

synchronized with the vibrating mirror, d_{328} and $d_{328} - d_{354}$ are recorded by a cathode ray oscillograph, and the equilibrium proportions and the reaction rate are determined (time of half completion $\approx 0\cdot1$ second).

Applications to 'tracer' methods

We have already seen how, in separating an impurity in cholesterol by fractional crystallization, the impurity can be traced in the various fractions by its absorption spectrum (page 106). Allsopp and Szigeti (1946) have used similar methods to separate two photochemical products of the irradiation of 3 : 4-benzpyrene. The irradiated product is soluble in water and shows absorption bands at 3600 and 2760 Å. It was found that, after making the aqueous extract alkaline and shaking it with ether, the ether component showed the band at 2760 Å only, the aqueous layer showing the 3600 Å band unchanged. The 2760 Å band was, however, considerably reduced in intensity. Thus the absorption spectra showed that the photochemical product consists of two substances of which one may be separated by solution in ether.

In attempts to synthesize natural substances the identity of the synthetic and natural product may be inferred when the chemical and physical properties of the two are identical. Among physical properties to be tested are the melting point, boiling point, solubility—and absorption spectra. The use of isotopic carbon

(C^{13}) as a tracer element is important in following natural processes. A spectrophotometric method of estimating $C^{13}O_2$ is
described on page 137.

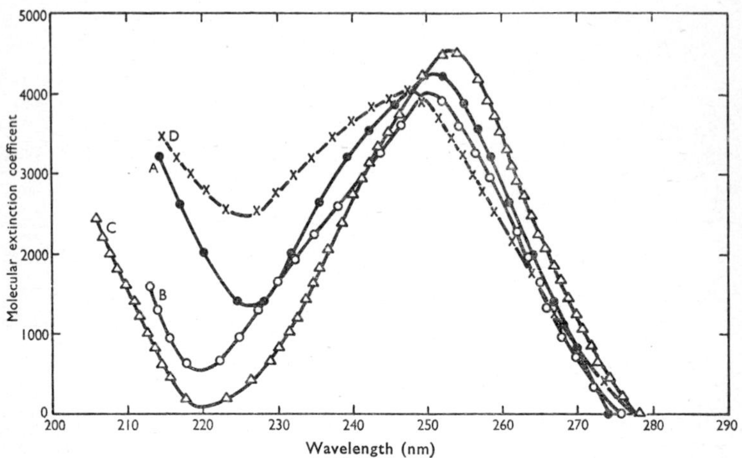

FIG. 6.4. Absorption spectra of basic cleavage product of aneurin hydrochloride (*A*), together with the spectra of substituted thiazoles (4-methyl-
thiazole hydrochloride (*B*), 2 : 4-dimethylthiazole hydrochloride (*C*), and
2-hydroxy-4-methylthiazole hydrochloride (*D*). (From Morton, 1942.)

Structure of vitamin B₁

Absorption data have proved extremely useful in synthesizing
and establishing the structure of a number of vitamins, and
Morton (1942) gives many examples, such as the brilliant and con-
clusive use of spectroscopic data by Williams and other workers
for determining the structure of vitamin B_1 (aneurin). This is
well summarized in Morton and the brief comments below do not
do justice to the beauty of the deductive processes used. The
empirical formula of aneurin hydrochloride was found to be
$C_{12}H_{16}N_4OS.2HCl$. From this can be prepared a basic cleavage
product with empirical formula C_6H_8NSCl, and Fig. 6.4 shows
the similarity (both in wavelengths of maxima and minima and
in the molecular extinction coefficients) of its absorption spectrum
to certain substituted thiazoles. Another cleavage product,
$C_6H_{10}N_4$, is obtained by treatment of aneurin with liquid ammonia;

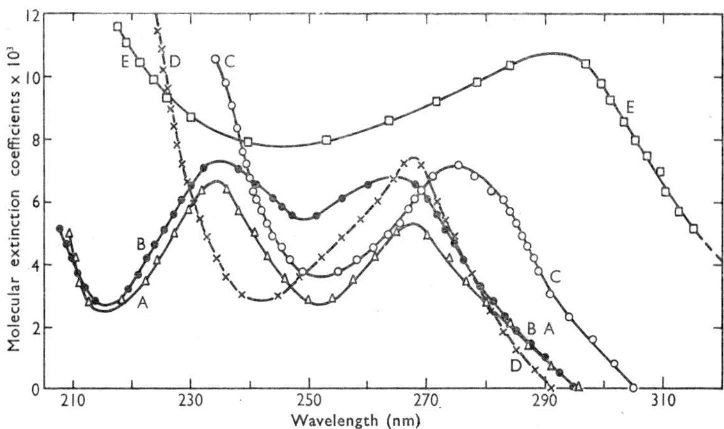

FIG. 6.5. Spectrum of liquid cleavage product of vitamin B_1 (*A*) compared with the spectra of 2 : 5-dimethyl-6-aminopyrimidine (*B*), 4 : 5-dimethyl-2 : 6-diaminopyrimidine (*C*), 5-ethyl-4 : 6-diaminopyrimidine (*D*), and 4-methyl-5 : 6-diaminopyrimidine (*E*). (From Morton, 1942.)

its spectrum is compared with several diamino compounds (Fig. 6.5), and the resemblance to 2 : 5-dimethyl-6-amino-pyrimidine is strong.

INFRA-RED SPECTRA

Before considering how a knowledge of infra-red spectra may be applied to structural chemistry, it will be necessary to discuss in more detail than in Chapter 2 the rotational and vibrational quantum jumps that are the source of the vibration-rotation absorption bands in the near infra-red.

Let us first consider changes of rotational energy only, which give rise to spectra in the far infra-red and microwave regions. If δE_{rot} is the change in rotational energy, then the frequency absorbed is given (where h is Planck's constant) by

$$h\nu = \delta E_{rot} \qquad (6.1)$$

It can be shown that

$$E_{rot} = h^2 J(J+1)/8\pi^2 I \qquad (6.2)$$

where I is the moment of inertia of the molecule about the axis of rotation, and J is the rotational quantum number. In this

H

expression J is the only variable, and the quantum theory shows that it may change either not at all or by a single unit, i.e.

$$\delta J = +1, 0, \text{ or } -1$$

The lines making up the rotational spectrum will therefore have frequencies given, for $(J-1) \rightarrow J$, by

$$h\nu_{\text{rot}} = h^2\,[(J+1)J - J(J-1)]/8\pi^2 I = 2Jh^2/8\pi^2 I$$

or
$$\nu_{\text{rot}} = Jh/4\pi^2 I \tag{6.3}$$

Since J can have values 0, 1, 2, ... etc., the lines will be separated by constant frequency intervals $h/4\pi^2 I$, so that the moment of inertia I, and thence the dimensions, of the molecule can be deduced directly from observations of this separation. Equation (6.3) shows that the frequencies of the rotational lines decrease as the moment of inertia of the molecule increases. For light molecules the rotational spectra lie in the far infra-red, e.g. for H_2O near $\lambda \leqslant 500\ \mu m$, and for heavier molecules in the microwave region, e.g. for $CHCl_3$ near $\lambda = 9$ cm and for RbBr near $\lambda = 20$ cm. Much work useful in considering structure is being done in this field but applications to chemical analysis have been limited. Since technique is different from that of optics, microwave absorption spectroscopy is outside the scope of this book, but the subject is well covered in a number of texts, including those of Townes and Schawlow (1955), Ingram (1955), and Walker and Straw (1961).

When both rotational and vibrational energies change simultaneously (*vibration-rotation* spectra of the near infra-red),

$$h\nu = \delta(E_{\text{rot}} + E_{\text{vib}}) \tag{6.4}$$

The quantum theory gives the vibrational energy of the molecule as

$$E_{\text{vib}} = (v + \tfrac{1}{2})h\nu_0,$$

where v is the vibrational quantum number ($v = 0$, 1, 2, etc.) and ν_0 the vibration frequency. Adding the rotational energy,

$$E_{\text{vib}} + E_{\text{rot}} = (v + \tfrac{1}{2})h\nu_0 + h^2 J(J+1)/8\pi^2 I$$

Thus, for a simultaneous transition from the vibrational level v_1 to v_2 and the rotational transition J_1 to J_2,

$$\delta E = h\nu_0(v_2 - v_1) + h^2\,[J_2(J_2+1) - J_1(J_1+1)]/8\pi^2 I$$

But, as before, $(\mathcal{J}_2 - \mathcal{J}_1)$ is generally restricted to the value 0 or ∓ 1 and for the '*fundamental*' vibration band, $v_2 - v_1 = 1$, so that we can write

$$\delta E = h v_0 \mp h^2 \mathcal{J}/4\pi^2 I$$

The frequencies absorbed are usually classified thus :

When $\mathcal{J}_2 - \mathcal{J}_1 = 0$ $v = v_0$ (Q)

When $\mathcal{J}_2 - \mathcal{J}_1 = 1$ $v = v_0 + h\mathcal{J}/4\pi^2 I$ (R)   (6.5)

When $\mathcal{J}_2 - \mathcal{J}_1 = -1$ $v = v_0 - h\mathcal{J}/4\pi^2 I$ (P)

Thus a band may have three branches—Q the central branch, R the positive branch on the higher frequency side, and P the negative branch on the lower frequency side.

The three equations above suggest that a vibration-rotation band should consist of a single line (Q branch) with a series of equally spaced lines on each side (P and R branches). But the moment of inertia of the molecule becomes greater as the value of rotational quantum number $\mathcal{J}$ increases and this can cause the lines of the P and R branches to lose their regular spacing. (This effect is well set out by means of a Fortrat diagram—see bibliography for textbooks on quantum theory.)

Bohr's correspondence principle

Although the frequencies in a vibration-rotation band are determined by the laws of quantum theory, ordinary classical considerations may be used to determine which of the various possible changes of vibration and rotation of a particular molecule are active in causing infra-red absorption, and to estimate the relative intensities of different absorption bands. This takes us back to the correspondence principle enunciated long ago by Bohr. The subject is discussed fully and clearly by Thompson (1944) and in various books on quantum theory. Here we shall merely demonstrate, by one or two examples, how the principle is applied:

1. In a linear non-polar molecule (O_2, N_2, etc.) which has no electric dipole moment the electric field of an electromagnetic wave cannot affect the vibration of the molecule; and thus O_2 and N_2 show no vibration spectra. Likewise, the linear

molecule CO_2 shows no vibration spectrum due to the symmetrical valency vibration

$$O{=}C{=}O \rightleftharpoons O{=\!=\!=}C{=\!=\!=}O$$

CO_2 does however exhibit vibration spectra due to the two other possible modes of vibration—the perpendicular vibration illustrated on page 117 and an asymmetric valency vibration.

2. In certain cases the electric field of an electromagnetic wave that affects the vibration of a molecule *must* also interact with the ions to affect the molecular rotation. For this reason, in certain vibration–rotation spectra, the Q branch is missing. An example of this is seen in the vibration–rotation spectrum of HCl (Fig. 6.6). (Note that in this spectrum each rotational line is double, because there are two types of molecule present, HCl^{35} and HCl^{37}.)

Intensity distribution in a band

In the undisturbed molecule, at absolute temperature T, the most probable rotational level will be that giving the molecule rotational energy equal to $\frac{1}{2}kT$, where k is Boltzmann's constant. The most probable change of rotational level will therefore be a

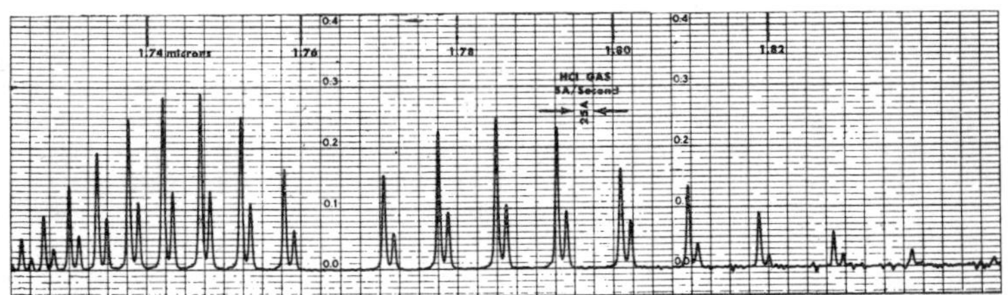

FIG. 6.6. Absorption of HCl gas—the first overtone of the vibration-rotation band. A recording obtained from a Cary spectrophotometer as illustrated on page 179. Note that the Q branch is missing.

change of quantum number of unity from this particular value; from equations (6.2) and (6.5) may be deduced the value of ν_{rot} showing maximum absorption, namely $\sqrt{(kT/2\pi^2 I)}$. The positions of these maxima are clearly seen in Fig. 6.6.

The vibration–rotation band of ammonia

The vibration–rotation band of ammonia gas at 10 μ recorded in Plate 4 has a considerably more complex structure. The NH_3 molecule is a pyramid with the N atom at the apex of the pyramid, and can undergo inversion when the N atom moves through the plane containing the three H atoms. Because of the resonance between the equilibrium positions on each side of this plane, the energy levels are split into two. Thus the whole band consists of two interlaced bands each having P, Q and R branches; the centres of the two Q branches are at 932 and 968 cm⁻¹. In addition, each line of the band is multiple because there are two axes about which the molecule can rotate—parallel and perpendicular to the axis of symmetry.

Position of vibration-rotation bands

Bands in the near infra-red spectrum may in general be ascribed to relative vibration of two adjacent parts of a molecule. In a molecule consisting of n atoms, each atom has three possible directions of motion, so that there are $3n$ degrees of freedom for the whole molecule. Of these, three account for translational and three for rotational motions of the whole molecule, leaving $3n - 6$ possible modes of vibration. Such vibrations may be along the lines joining the adjacent parts (parallel or valency vibrations) or perpendicular to the valency bonds (perpendicular vibrations). Perpendicular vibration is signified by vertical arrows, as indicated in the following representation of a linear symmetrical molecule.

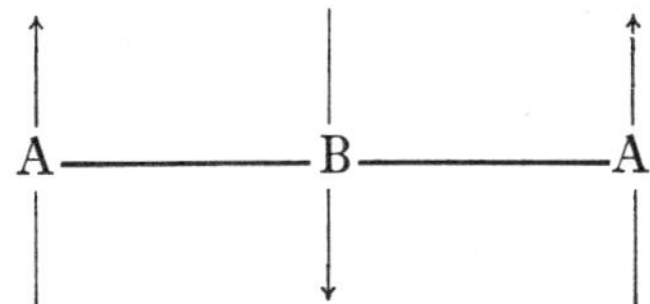

There are also, especially in complex molecules, other modes of vibration that it is not possible to describe in a short phrase.

The vibration of a carbon and hydrogen atom along the valency bond joining them gives rise to an infra-red frequency of about 2900 cm⁻¹ ($\lambda \sim 3\cdot4 \,\mu$m). Since the mass of the carbon atom is much greater than that of the hydrogen atom, the masses of the other atoms attached to the carbon atom have only a small effect on the C—H frequency (nevertheless the effect has been investigated and provides the basis of a useful analytical method—see below).

For heavier atoms one would expect the frequency of vibration to be less; for a C—C bond, for example, $\bar{\nu} = 990$ cm^{-1} ($\lambda \sim 10$ μm). For multiple bonds frequencies are larger; C=C gives $\bar{\nu} \sim 1620$

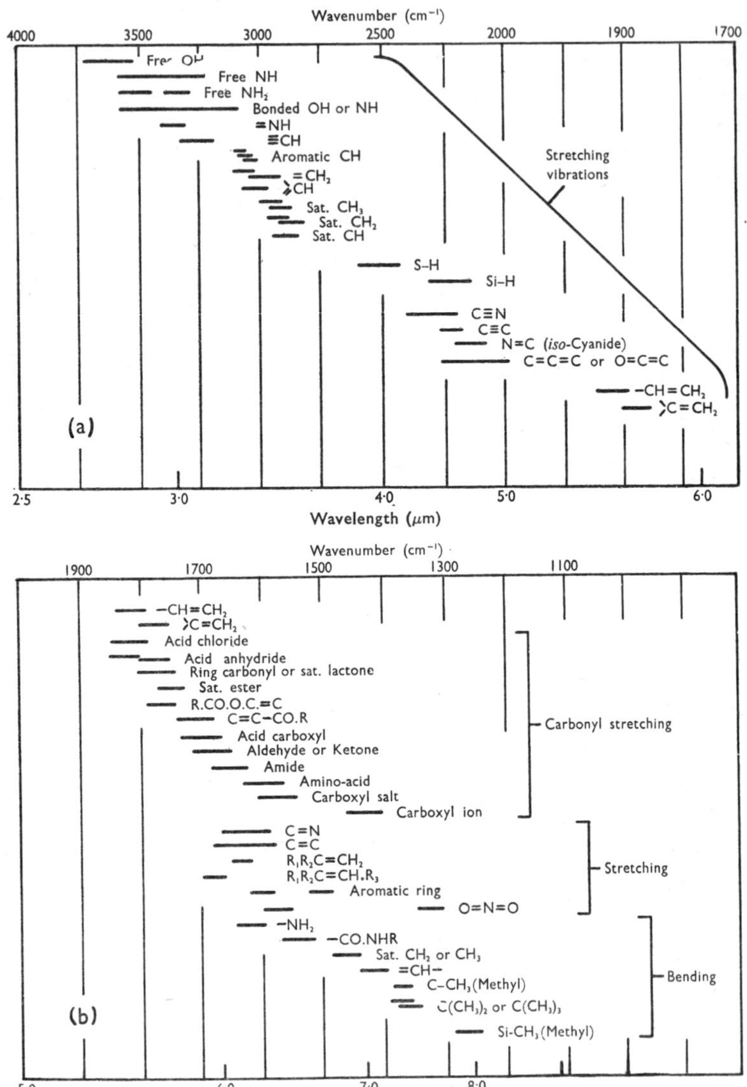

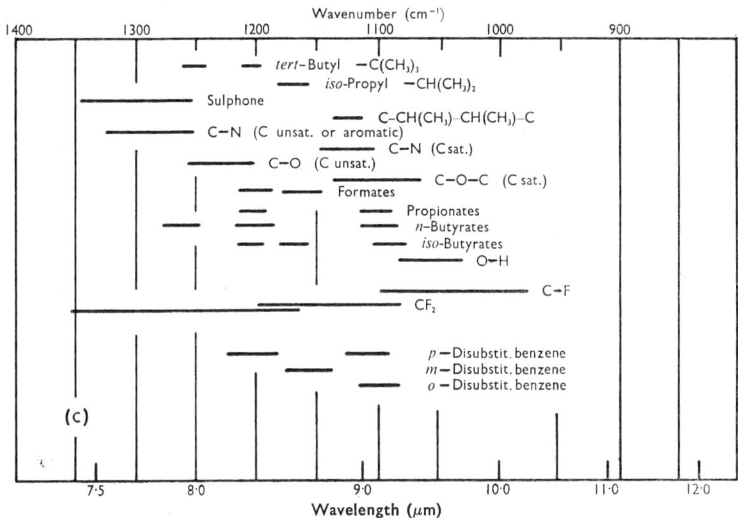

FIG. 6.7.   Positions of vibrational bands characteristic of certain bonds or groups. (From Price, 1955.)

cm^{-1} ($\lambda \sim 6 \cdot 2\,\mu$m), and C≡C gives $\bar{\nu} = 2100$ cm^{-1} ($\lambda = 4 \cdot 8\,\mu$m). Perpendicular or twisting vibration in general gives rise to smaller restoring forces, and so smaller frequencies; C—H twisting vibration gives a band near $\bar{\nu} = 1400$ cm^{-1} ($\lambda \sim 7\,\mu$m).

Fig. 6.7 summarizes some frequencies typical of certain groupings. Such data are valuable in diagnosing structure but must be used with caution and careful consideration of the disturbing effect of neighbouring groups. The subject is discussed more fully by Barnes, Gore, Liddell and Williams (1944), Bellamy (1954) and Freeman (1965).

Overtones and combination of vibrational frequencies

Overtones of a vibrational frequency may occur in the absorption spectrum. The fundamental vibration of HCl is at 2900 cm^{-1}, and the band shown in Fig. 6.6 is the first overtone. The valency vibration of the OH group is at about 3600 cm^{-1} ; first and second overtones, with frequencies about twice and three times the fundamental, may also be present, occurring at 7050 and 10,320 cm^{-1} for phenol in carbon tetrachloride.

Combinations of two vibrations are possible. The benzene nucleus, for example, has vibration frequencies at 993, 1010, 1485 and 1595 cm^{-1}. In phenol, one or two quanta of these nuclear vibrations (ν_n) may be combined with the OH valency vibration (ν_{OH}), or one of its overtones; there are thus absorption bands in the spectrum with frequencies in the regions

$$\nu_{OH}+\nu_n, \ \nu_{OH}+2\nu_n, \ 2\nu_{OH}+\nu_n, \ 2\nu_{OH}+2\nu_n, \ 3\nu_{OH}+\nu_n, \ 3\nu_{OH}+2\nu_n$$

It is convenient to examine overtones and combinations because shorter wavelengths are more accessible; a quartz, or even glass, monochromator may perhaps be used in place of rocksalt, and the greater energy available also allows a smaller instrument and less sensitive detecting arrangements to be used. We can, for instance, use the frequencies ($3\nu_{OH}+\nu_n$) above to observe the benzene nuclear vibrations (which occur as fundamental bands in the region 6–10 μm) at wavelengths near 1 μm with glass optics.

A modified combination band involves simultaneous transitions in two molecules, so that the energy of an absorbed photon is divided between them. Thus Ketelaar and Hooge (1955) find that various substances (PCl_3, S_2Cl_2, etc.) dissolved in CS_2 give absorption bands that have shifted to wavenumbers higher by 1515 cm^{-1}, which corresponds to an absorption band of CS_2. This again transfers the low-frequency absorptions of such heavy molecules to a region more readily observable.

SOME TYPICAL APPLICATIONS

Rotational spectrum and moment of inertia of N_2O

Using a grating spectrophotometer not unlike that of Fig. 9.7, Palik and Rao (1956) have observed in the far infra-red a large number of lines of the pure rotation spectrum of nitrous oxide— from $\bar{\nu}=16\cdot74$ to $\bar{\nu}=34\cdot34$ cm^{-1}, corresponding to $\mathcal{J}=19$ to 40 in equations (6.3). They used a modified equation

$$\bar{\nu}=2B\mathcal{J}-4D\mathcal{J}^3$$

where $B=h/8\pi^2cI$, and where the final term allows for the increase of moment of inertia with angular velocity, because of centrifugal forces; D is the 'centrifugal stretching coefficient'.

Values of B and D are calculated from the measurements:

$B = 0.41901$ cm^{-1} and $D = 0.19 \cdot 10^{-6}$ cm^{-1}. From B it is easily calculated that the moment of inertia of the molecule is $I = 2.06 \cdot 10^{-38}$ g cm^2.

The allyl halides

A good example of the way in which these principles can be applied is an analysis of the absorption spectra of three allyl halides that has been carried out by Thompson and Torkington (1946). Fig. 6.8 is reproduced from their paper. There are nine atoms in the molecule (CH_2:CH—$CH_2.X$) and so there are 21 (i.e. $3n - 6$) modes of vibration. Thompson and Torkington are able to explain all the bands observed in terms of the fundamentals, overtones, and combinations of 21 fundamental vibrations. The frequencies near 3000 cm^{-1}, for example, are obviously to be attributed to C—H valency vibrations. The vibration ranging from 890 cm^{-1} to 825 cm^{-1} in the different compounds is probably the C—C valency and the frequency 1640–1645 cm^{-1} that occurs in all the curves is due to C=C valency vibration. The vibration at 594 cm^{-1} for chloride, at 541 cm^{-1} for bromide, and at 498 cm^{-1} for iodide, is probably due to C—halide valency vibration. In sorting out these frequencies, use is also made of the Raman spectra of the substances.

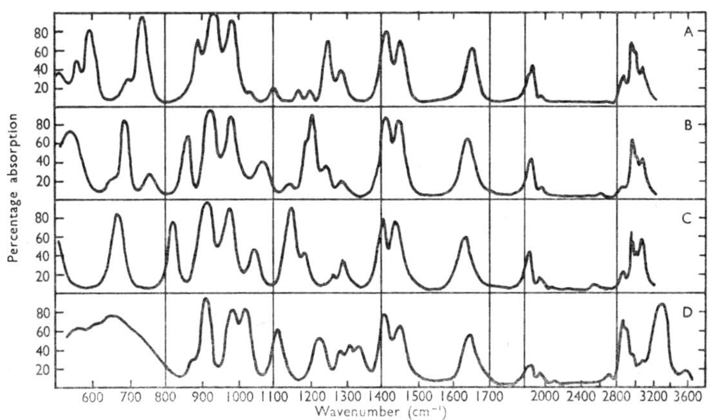

FIG. 6.8. Absorption spectra of allylchloride (A), allylbromide (B), allyliodide (C), and allylalcohol (D). (Thompson and Torkington, 1946.)

Molecular association of hydroxyl compounds ; the hydrogen bond

Vapours and dilute solutions of compounds with an —OH group all show an absorption band near 2·75 μm ($\bar{\nu}=3640$ cm^{-1}) that is attributed to valency vibration of the O—H bond. It was first shown by Errera and Mollet (1936) that, whereas a dilute solution of ethyl alcohol in carbon tetrachloride shows this band only (concentration not more than 0·02 mol/litre), on *increasing* the concentration the intensity of this band *decreases* and a new broad band at lower frequency appears centred about $\nu=3300$–3500 cm^{-1}.   This band is attributed to the presence of associated molecules. Errera, Gaspart and Sack (1940) have investigated solutions of ethyl alcohol in carbon tetrachloride. Measurements on very dilute solutions where the degree of association is negligible serve to determine the molecular extinction coefficient of the monomer, and measurements on the pure alcohol give the specific extinction coefficient of the polymer (one cannot refer to the molecular extinction coefficient in this case until one knows the molecular weight of the polymer). One might expect that it would then be possible, by application of Beer's law, to determine the amounts of polymer and monomer at any given concentration. If this were so, one could then apply the law of mass action to determine the number of monomer molecules in the polymer— i.e. one could determine n in the equilibrium

$$n\ \mathrm{C_2H_5OH} \rightleftharpoons (\mathrm{C_2H_5OH})_n$$

The results, however, show that association takes place in two stages. Fox and Martin (1940) have been able to apply the law of mass action to similar measurements on benzyl alcohol solutions with results that suggest the equilibrium

$$4\ \mathrm{BzOH} \rightleftharpoons 2\ (\mathrm{BzOH})_2 \rightleftharpoons (\mathrm{BzOH})_4$$

Fig. 6.9*a* shows the absorption of dilute solutions of benzyl alcohol in carbon tetrachloride, the peaks at 2·75 and 2·765 μm being attributed to two different forms of monomeric benzyl alcohol. Fig. 6.9*b*, however, showing the absorption of a more concentrated solution and of pure liquid benzyl alcohol, has peaks at 3·0 μm due to *inter*molecular association.

This form of association is generally attributed to the ' hydrogen

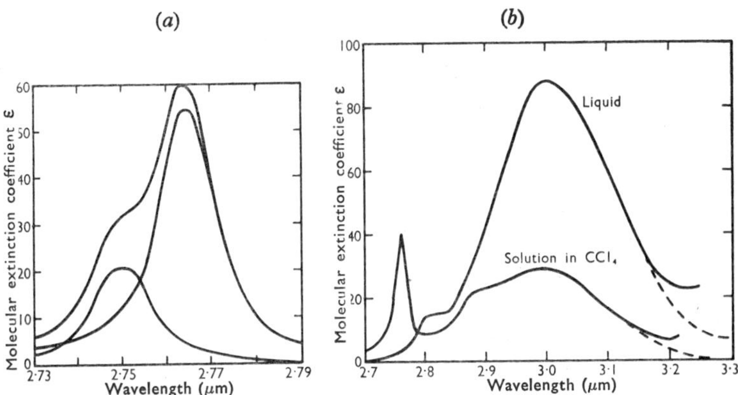

FIG. 6.9. Absorption curves of benzyl alcohol. (a) Dilute solutions (0·0005–0·03 mol/litre). (b) More concentrated solution (0·17 mol/litre) and pure benzyl alcohol. (From Fox and Martin, 1940.)

bond', which shows itself as an attractive force between a hydrogen atom and an oxygen atom of a neighbouring formation, i.e., in the above example, an oxygen atom in a neighbouring alcohol molecule.

A somewhat different form of hydrogen bonding occurs *intra*molecularly in methyl salicylate, as indicated by the structural formula

The absorption curve of this compound (Fig. 6.10) shows no trace of the normal OH band at 2·77 μm but shows instead the broad band at longer wavelength (3·12 μm) characteristic of association. Since the hydrogen bonding in this case is self-contained in each molecule, one would expect the molecular extinction coefficient to be unaffected by the concentration, and this is found to be so.

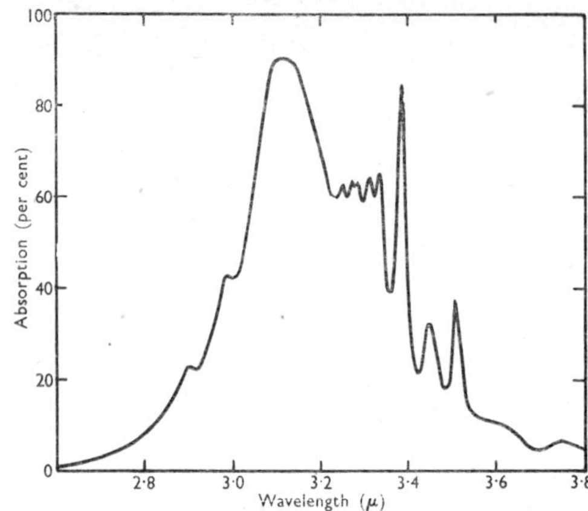

FIG. 6.10. Absorption spectrum of methyl salicylate. The band at 3·12 μm is attributed to intramolecular association. (From Fox and Martin, 1940.)

Deuterium substituted Compounds

If deuterium is substituted for hydrogen at one point in a molecule, the binding forces remain unchanged but the moment of inertia is increased. Such a substitution will lower the frequency of any vibration involving this particular part of the molecule.

Replacing hydrogen by its isotope doubles the mass and so has a much bigger effect than the isotope substitution of other atoms. Considerable use has been made of this method for assigning the different frequencies of a particular spectrum to the correct vibrations. Halverson (1947) gives a fairly detailed account of the many uses of this method. Note also Fig. 10.2 (page 213) showing absorption curves for H_2O and D_2O.

ABSORPTION OF POLARIZED RADIATION

Dichroism (pleochroism)

It has been explained (page 115) that emission or absorption of radiation occurs only when the transition (electronic, vibrational, or rotational) involves a change of electric moment.* It follows that the electric field (also called the electric vector) of the ab-

---

* In this elementary discussion, we ignore the weaker transitions known as magnetic dipole, etc.

sorbed, or emitted, electromagnetic radiation is in the same direction as this change of electric moment, or transition moment. In a substance in which the molecules are regularly arranged, as in a crystal or a polymer, the amount of absorption will depend on the relative orientation of the molecules and the electric vector, i.e. on the relation between the plane of polarization* of the radiation and the crystal axes.

This phenomenon, named pleochroism or dichroism, has long been known. The production of plane polarized light by transmission of a beam of unpolarized light through tourmaline or polaroid are common examples in which there is intense absorption for one particular direction of the electric vector. Dichroism is of considerable use in the investigation of crystal structure and in the assignment of absorption bands to particular transitions. For example, L. H. Jones (1953) found (see Fig. 6.11) that a crystal of $KAu(CN)_2$ shows two infra-red absorption bands when the electric vector is parallel to the c-axis of the crystal, and that these bands are much reduced in intensity when the electric vector is perpendicular to this axis. The conclusions that may be reached from these measurements were later considered more fully by L. H. Jones (1954).

Infra-red dichroism has also been used by Fraser and Price (1952) and by Elliott (1953) to investigate the structure of polypeptide chains. The former workers point out that great care must be taken in interpreting experimental results because the transition moment, and hence the electric vector of the absorbed beam, may not lie exactly along a bond; the phenomenon of electron following (movement of electrons during a vibration in favour of one of two alternative resonating structures) may incline the transition moment to the valency bond. Fraser and Price conclude that for the CO valency vibration observed for some peptides at 1650 cm^{-1} this inclination may be about 15 degrees. Fraser (1956) has shown how measurements of the dichroic

---

* It must be remembered that the electric vector or electric field in an electromagnetic wave is perpendicular to the 'plane of polarization', as defined in the older textbooks on optics. To avoid ambiguity it is better not to use this term and to state the direction of the electric vector. Thus, in a plane polarized beam obtained by reflection at the Brewster angle, the electric vector is perpendicular to the plane containing the incident and reflected beams and the normal to the surface, i.e. it is perpendicular to the plane of reflection.

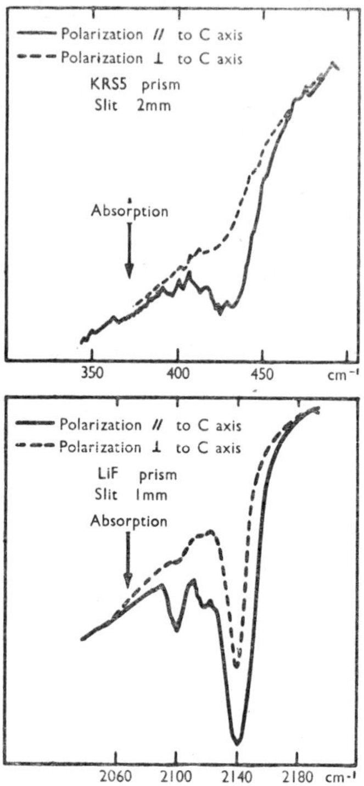

FIG. 6.11. Absorption of polarized radiation by monocrystalline $KAu(CN)_2$. The incident radiation is plane polarized with the electric vector parallel (*//*) and perpendicular (⊥) to the c-axis of the crystal.
(From L. H. Jones, 1953.)

ratio in materials such as wool containing polypeptide chains can be used to calculate the fraction of oriented chains.

In the ultra-violet, measurements on a crystal of *p*-dimethoxy-benzene by Albrecht and Simpson (1955) have shown that, if the radiation is transmitted in a direction perpendicular to the plane of the molecule, an absorption band occurs at 2300 Å when the electric vector of the radiation is parallel to the line of the *para* substituents. Another band occurs at 2900 Å when the electric vector is perpendicular to this line. Conclusions that the excited states of the benzene molecule give rise to bands at 2000 and 2600 Å are based on these findings.

Circular dichroism

As shown in textbooks of optics, a beam of plane polarized light, in which the alternating electric field is confined to one plane, is equivalent to two beams of opposite circular polarizations; i.e. in which the direction of the electric field rotates in opposite directions about the direction of propagation. Looking against the direction of the beam, the one in which the electric field is rotating clockwise is said to have right-handed or dextro-circular polarization, and the beam in which the electric vector is rotating anti-clockwise has left-handed or laevo-circular polarization; dextro- and laevo- are abbreviated d- and l-.

Fresnel pointed out that optical activity (rotation of plane of polarization) is equivalent to the propagation of two beams with opposite circular polarizations with different velocities. In other words, optically active media have different refractive indices for the two directions of circular polarization. But the occurrence of a refractive index different from unity is bound up with the existence of an absorption band in some part of the spectrum not too distant (see textbooks on electromagnetic theory). It follows that materials which are optically active and which therefore have different refractive indices for d- and l-circular polarizations must have different absorptions for d- and l-polarizations at wavelengths where they are absorbing. This is the phenomenon of circular dichroism; it was observed by Cotton in 1895 for solutions of copper tartrate and chromium tartrate, and is often called the Cotton effect.

Circular dichroism may be expressed quantitatively at some wavelength as the molar circular dichroism, the difference $\Delta\epsilon$ of the molecular extinction coefficients for l- and d-circular polarizations; i.e.

$$\Delta\epsilon = \epsilon_l - \epsilon_d$$

$\Delta\epsilon$ may be positive or negative. The relative value defined as

$$g' = \Delta\epsilon / \tfrac{1}{2}(\epsilon_l + \epsilon_d) \tag{6.6}$$

is named the disymmetry factor. The value of g' rarely exceeds 0·2. The value of g' obviously varies with wavelength through the absorption band. A factor g is also used which is defined similarly, but using the areas ($\int\epsilon d\nu$) under the absorption bands instead of single values of ϵ.

Apparatus for measuring circular dichroism is described in Chapter 9. It is convenient to mention here briefly the related phenomenon of anomalous rotatory dispersion. Optical rotation of plane polarized radiation changes with wavelength and, in

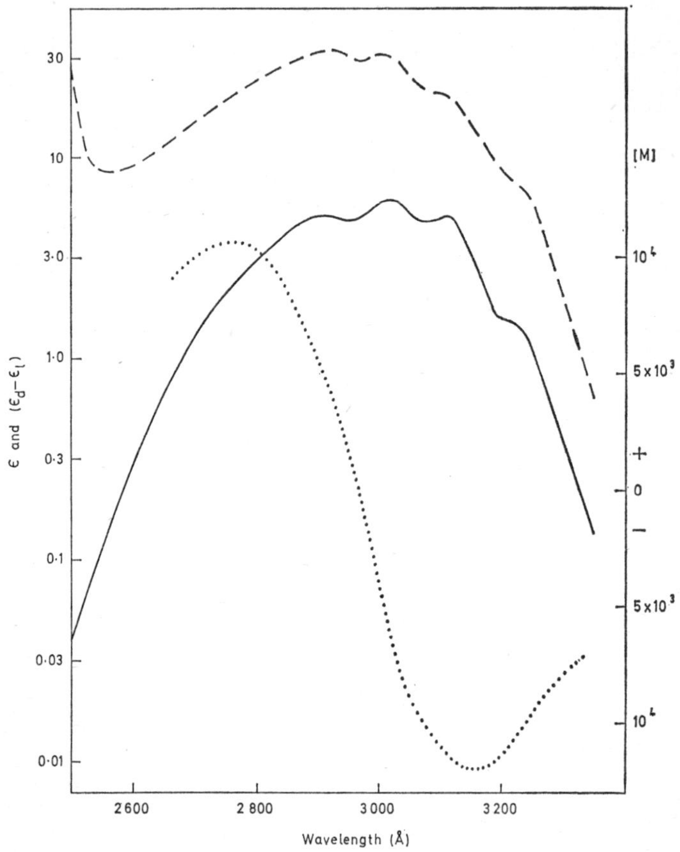

FIG. 6.12. Properties of a methanol solution of 3β-hydroxy-5α-androstan-16-one. - - - - unpolarized absorption spectrum. ——— circular dichroism. . . . . optical rotatory dispersion. The right-hand ordinates M apply to the last mentioned curve. M, the molar rotation = (molecular weight × specific rotation)/100.
(Reproduced from Mason, S. F. 1962.)

passing through a wavelength region showing circular dichroism, changes in an anomalous fashion as seen in Fig. 6.12. Thus measurement of this anomalous rotatory dispersion with a spectro-polarimeter provides a further means of characterizing the same phenomenon. But it must be remembered that the rotatory dispersion at a wavelength showing circular dichroism is superposed on a background of rotatory dispersion due to circular dichroism bands *elsewhere* in the spectrum. Analysis of rotatory dispersion data may therefore be complex and not always un-ambiguous; and so circular dichroism measurements are generally more reliable for estimations of rotational strength.

The subject is reviewed by Mason (1963) and the following data quoted from his paper form examples of typically large and small values of circular dichroism:

Compound	λ_{max}	ϵ_{max}	$\varDelta\epsilon$	g	chromophore
3β-hydroxy-5α-androstan-16-one	3000Å	30	$-5\cdot95$	$0\cdot17$	$\diagdown\text{C}{=}\text{O}$
α-bromopropionic acid	2300Å	400	$+2\cdot6$	$0\cdot006$	$-\text{C}-\text{Br}$

The properties of the first of these compounds are shown in Fig. 6.12, which shows at 3000 Å (the wavelength characteristic of the weak C=O chromophore—see Table 6.1), the total absorption $(-\,-\,-)$, the circular dichroism $(\underline{\hspace{1.5em}})$, and the anomalous rotatory dispersion $(\ldots)$. It is seen that at the wavelength of ϵ_{max} ($\sim$3000 Å), $g'=\varDelta\epsilon/\epsilon\approx4/30\approx0\cdot13$, whereas according to the tabulated data above the ratio of areas under the bands is $g=0\cdot17$.

The low value of $\epsilon_{max}\approx30$, indicates that the electronic transition involved is forbidden as an electric dipole transition.

INTENSITY OF AN ABSORPTION BAND

OSCILLATOR STRENGTH

The magnitude of a transition is important as well as its position and expressions relating the intensity of an absorption band to the value of the transition moment (p. 125) have been developed. The relationship has been discussed in several books dealing with the quantum theory. It can be shown that the total intensity of an absorption band, already defined by equation (3.29) on page 53, may be given by

$$B = \int \mu \, d\bar{\nu} = \frac{f \pi N_A e^2}{mc^2} \tag{6.7}$$

where N_A is Avogadro's number, e and m are the charge (expressed in electrostatic units) and mass of the electron, and c is the velocity of light in vacuum. This equation readily reduces to the useful numerical relation:

$$f = 4 \cdot 32 \, . \, 10^{-9} \int \epsilon \, d\bar{\nu}$$

where ϵ is the molecular extinction coefficient.

In quantum theory, f is the probability of the transition and may be unity for a fully allowed transition, or it may be less than unity if the transition is not fully allowed or if there are several possible transitions from a given ground state. It is referred to as the oscillator strength of a transition, or simply as the 'f factor'. If f is taken as the number of oscillators per atom giving rise to the band, the above equation may also be deduced from classical theory.

Total band intensity, or oscillator strength, has been investigated for a number of molecular types by Mulliken (1939), who has published an important series of papers. The calculations performed by Larnaudie (1952) for the intensities of the vibrational transitions in the infra-red spectrum of *cyclo*hexane are an example of practical application. Correlation of calculated and observed intensities has been of use in identifying lines whose origins were otherwise in doubt. Mills and Thompson (1955) have used the theory the other way round to calculate, from the intensities of vibrational bands of dimethylacetylene as measured by the methods described on page 54, the dipole moment associated with some of the bonds. They conclude that the charge distribution in the molecule is such that there is probably a net positive charge on the H atoms.

The f-value of the absorption band giving rise to the colour induced in the visible spectrum of KCl by excess K has been determined by Rauch and Heer (1957). They determined the total intensity of the band by measuring the area under the curve and for the same specimen estimated the amount of excess K; the value so found is $f = 0.66$.

CHAPTER 7

Analytical Applications

METALLURGICAL ANALYSIS

A WIDE range of methods has been developed over recent years for estimating metals from the absorption or fluorescence spectra of solutions of salts or complexes.

The simplest methods apply to solutions that may contain several metals, but only one as a coloured compound. When dealing, for example, with a copper alloy containing, say, manganese, phosphorus, nickel and iron, it is possible to add sodium thiocyanate to the solution to form red ferric thiocyanate, whose absorption in the green may be measured; other compounds in the solution have negligible absorptions in this wavelength region. In some cases, when two coloured compounds are present, it is possible to measure both. For example, a steel containing both chromium and manganese may be dissolved and oxidized by suitable reagents to yield dichromate and permanganate salts. Both these are coloured but have absorptions in different spectral regions, violet and yellow-green respectively. Vaughan (1941 and 1942) and Haywood and Wood (1957) have shown how, using abridged spectrophotometry, to estimate the two compounds in the presence of each other by measuring their absorption with the violet and yellow mercury lines.

The use of metal complexes

As pointed out by West (1966) one method of increasing the sensitivity of a method, i.e. of increasing the probability of absorption of radiation, is ' to spread a mesh of closely packed π-orbitals in the molecule to capture photons and secure a transition. A method can be made relatively specific by choosing a close-packed complexing molecule which will admit only small ions. A complex of silver with 1:10-phenanthroline and bromopyrogallol red gives $\epsilon = 51\ 000$ which permits estimation down to $M/(50 . 10^6)$, taking the limiting precision in absorbance as 0·001 and using a 1 cm cell.

Estimation of phosphorus

Jurkin, Kirkbright and West (1966) describe a sensitive method for phosphorus. The first stage is conversion to phospho-molybdic acid $[H_3PO_4(MoO_3)_{12}]$; the twelve (MoO_3) groups associated with one P atom give effectively a twelvefold amplification of the sensitivity for P. The phosphomolybdic acid then forms a green coloured complex (absorption maximum at $\lambda = 7100$ Å) with 2-amino-4-chloro-benzenethiol, which has $\epsilon_{mol} = 359\,000$. Thus one can determine $0.2\ \mu g$.

Spectrofluorimetric estimation of magnesium

Magnesium forms a fluorescing complex with NN'-bis-salicylidene-2:3-diaminobenzofuran (SABF). Large metallic ions cannot penetrate the rigid structure of this molecule so that the number of ions which can interfere by forming similar complexes is limited. Calcium does form an absorbing, non-fluorescing complex, however, so that excess of calcium gives a low result for Mg (see theory on page 56, Chapter 3). But even a 200-fold excess of Ca can be tolerated, if another complex is added (strontium with ethylenediaminetetra-acetic acid, Sr-EDTA); the presence of this latter complex prevents extraction of the Ca-SABF complex. These theoretical aspects are discussed by Dagnall, Smith and West (1966), who also (1967) give experimental details; the fluorescence spectrum has a peak at $\lambda = 5450$ Å and is excited by irradiation at $\lambda = 4750$ Å. The instrument used is similar to the one illustrated in Fig. 9.12. They used the method for estimating Mg in water over the range $1–50\ \mu g/l$; and in blood plasma at a concentration of $20\ \mu g/l$. They estimate the detection limit at 10^{-3} ppm.

Estimation of beryllium by a spectrofluorimetric method

Kirkbright, West and Woodward (1965) make use of the fluorescent complex of Be with 2-hydroxy-3-naphthoic acid. The fluorescence is excited by $\lambda = 3800$ Å and measured at $\lambda = 4600$ Å. The fluorescence of the test solution is compared with that of a standard quinine sulphate solution (excitation at $\lambda = 3500$ Å, observation at $\lambda = 4500$ Å). The precision is estimated to be $\mp 0.003\ \mu g$ Be. Interference by the fluorescent complex that would be formed with Al is eliminated by masking with the Ca-complex of CDTA (trans-1:2-diamino-cyclohexane tetra-acetic acid).

C, S and O in steel by the Luft method

Carbon and sulphur are burned to CO_2 and SO_2/SO_3 and determined by a 'Luft' apparatus similar to that shown in Fig. 9.15, page 197. The detector chambers C are filled with the appropriate gas.

Oxygen in steel can be determined by conversion to CO—burning with excess carbon in a graphite crucible. In this case, the Luft chambers C of Fig. 9.15 contain CO.

Molybdenum, titanium and vanadium

Molybdenum, titanium and vanadium all form yellow-red complexes with hydrogen peroxide, and each may be measured by ordinary absorptiometric methods when the others are absent. But the absorption curves, although they overlap, have peaks at

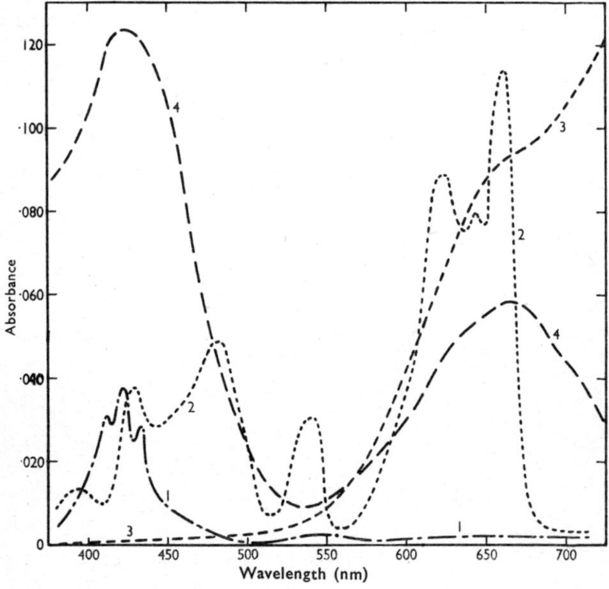

FIG. 7.1. Absorption curves of sulphates of uranium and vanadium (from Canning and Dixon, 1955). 1, U(vi), 2, U(iv), 3, V(iv), 4, V(iii).

considerably different wavelengths—3300 Å (Mo), 4100 Å (Ti) and 4600 Å (V). Weissler (1945) has shown that satisfactory estimation of these, together, may be made by measurement at these three wavelengths (calculation by method given on page 85).

An example of the manipulation sometimes necessary to eliminate the effects of absorbing impurities is the method worked out by Canning and Dixon (1955) for the estimation of uranium and vanadium in the presence of chromium, iron, titanium, and rare earths. Fig. 7.1 shows the absorption curves of solutions of salts of uranium and vanadium before and after reduction with ferrous sulphate. The change in absorbance on reduction (curve 2 minus curve 1) at 6500 Å is a measure of the amount of uranium present, and the change (curve 3 minus curve 4) at 7000 Å determines the amount of vanadium. The absorption bands overlap and this is therefore an example of the type of mixture dealt with on page 85.

ATOMIC ABSORPTION

This subject is properly a branch of absorption spectrophotometry. However, it is now a large subject in its own right, and there are textbooks devoted exclusively to it [e.g., Elwell and Gidley (1961)]. A short note on the principles must therefore suffice here.

Walsh (1955) suggested that the estimation of elements could be made by evaporating and decomposing compounds in a flame. The concentration of the element in the flame may then be determined by measuring the absorbance of the flame using a source emitting a resonance line of the element in question.

In the flame the lines emitted by the atoms are broadened to about 0·02 Å; for Beer's law to hold it is important to measure the absorption in a narrow range near the centre of this width. The source must therefore emit very narrow lines and it usually consists of a hollow-cathode discharge (freedom from Stark effect, low temperature, low pressure). The remainder of the equipment consists of a flame into which the specimen, often as a solution, is sprayed, a monochromator or narrow-band interference filter and a photocell.

If a specimen is so decomposed in a flame that all of the element to be determined exists in the flame as free, unionized atoms in the ground state, then the absorbance of the flame is proportional to the concentration in the sample. This assumption, while not completely valid, is much more nearly true than any analogous assumption about the conditions in emission spectrochemical analysis; and this is one of the advantages of the atomic absorption method. Nevertheless, the method still depends on calibration with known specimens. To mention two examples of interference:

1. The presence of certain anions (silicate, phosphate) in estimating strontium causes some of the latter to be present in the flame as undecomposed salts.

2. In the case of alkali and alkaline earth elements, whose ionization potentials are small, the atoms are easily ionized, thus reducing the concentration of ground-state atoms. In practice, this effect can be reduced by adding free electrons to the flame to encourage reversal of the ionization process; this may be done by adding an excess of another alkali. See Janauer, Mangan and Smith (1967) and Hermann and Alkemade (1963).

A further example of an estimation by atomic absorption is given on page 141 (pentachlorophenol).

OTHER INORGANIC ANALYSES

Nitrates

The nitrate ion shows an absorption maximum at 3020 Å and this band has been used for its estimation in the presence of nitrites and other anions. It was used by Dolance and Healy (1945) for estimation of nitrate in silver nitrate baths.

Spectrophotometric titration of inorganic cyanates

Spectrophotometry can be used as a means of locating the end point of a titration. Trusell, Argabright and McKenzie (1967) have used the method for titration of cyanate which forms complexes with cobalt ions having the formulae $Co(CNO)_x$ where x may be 2, 3 or 4. $Co(CNO)_4$ has an absorption maximum at $\lambda = 6420$ Å; addition of a cobalt salt solution thus shows an

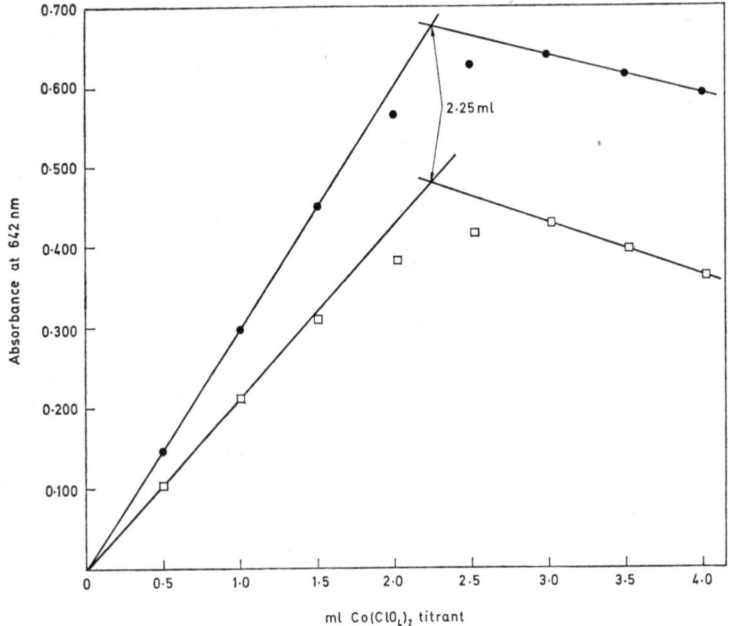

FIG. 7.2. Titration of cyanate against a cobalt salt to form complexes— $Co(CNO)_4$ and $Co(CNO)_3$.
(Reproduced from Trusell, Argabright and McKenzie, 1967.)

absorption maximum near the composition $Co(CNO)_4$, with decreasing absorption as more $Co(CNO)_3$ is formed. This is shown in the curves of Fig. 7.2; the method was tried in the two solvents there indicated. Dimethylformamide is regarded as the most satisfactory solvent for this method.

Isotopes of carbon

The effect of isotopic substitution on the positions of rotational and vibrational bands has already been discussed in connection with problems of molecular structure (page 124). Estimation of a particular isotope is also important in conducting 'tracer' experiments to follow the movement or reaction rate of an injected isotopic material. Absorption measurements can be applied in this way to determine the fraction of the C^{13} isotope in carbon dioxide. The vibration-rotation band near $4·4\,\mu m$ is such that the

P branch of $C^{13}O_2$ is entirely clear of the band of $C^{12}O_2$, so that estimation of the former may be made by a single measurement. The method has been worked out by Kluyver and Milatz (1953 and 1954), using a selective detector (infra-red gas analyser described on page 196). The sensitivity is 0·005 per cent of $C^{13}O_2$ in an 8-mg sample.

ESTIMATION OF HYDROCARBONS AND OTHER ORGANIC COMPOUNDS

Hydrocarbons

The spectrophotometric estimation of hydrocarbons was developed considerably during the 1939–45 war. Saturated paraffins show no absorption bands in the near ultra-violet spectrum and so their infra-red bands are used. Barnes, Gore, Liddell and Williams (1944) have shown the multiplicity of bands that are available for the 5–10 μm regions; their book contains 363 absorption curves covering that range (also a large bibliography). We have already discussed (pages 86–8) the methods used for hydrocarbons by Gordon and Powell (1945) and by Fry, Nusbaum and Randall (1946).

Aromatic hydrocarbons have absorption in the near ultra-violet as well as in the infra-red, so that either or both of the regions may be used according to their suitability for a particular application.

Inhibitors in polymers

It is common practice to add an inhibitor to plastic polymer materials to prevent degradation of the plastic by oxidation. The concentration of the inhibitor is therefore a guide to the stability of the material. Banes and Eby (1946) show how this may be estimated by dissolving the specimen in a suitable solvent and measuring the ultra-violet absorption in the region of the maximum. They correct for the background absorption due to the polymer by measuring at three wavelengths as described on page 95. They quote results for n-phenyl-2-naphthylamine used as an inhibitor in various synthetic rubbers, the wavelengths for these measurements being 2840, 3090 (λ_{max}) and 3320 Å. The method is found to be rapid and more reliable than chemical methods.

Acetone

The method used by Barthauer, F. V. Jones and Metler (1946) for estimating acetone is important in the production of propylene from isopropyl alcohol, when acetone is produced as a by-product and di-*iso*propyl ether is also present. Acetone has a broad absorption band centred near 3000 Å (see page 102) but the other substances have no appreciable absorption near this region. The spectrometric estimation was therefore straightforward ; 2 : 2 : 4-tri-methylpentane was used as a solvent, and tests showed that, with a photoelectric spectrophotometer and a waveband of 5 Å, Beer's law was obeyed.

Mixture of stereo-isomers of benzene hexachloride

The isomers have non-overlapping absorption bands at 762, 745, 845, and 772 cm^{-1} (see Marrison, 1949). A feature of the method is that the varying solubility of the isomers makes it necessary to take the measurement at 845 cm^{-1} with a specimen dissolved in nitromethane and the measurements at the other wavenumbers with a solution in methyl acetate. This analysis is important because the γ isomer is used as an insecticide.

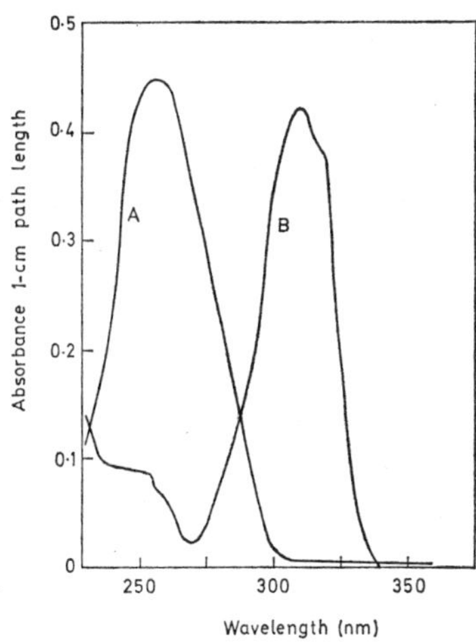

FIG. 7.3. Absorption curves in sodium acetate buffer solution. (*A*) Paraquat, 4 mg/l. (*B*) Diquat, 4 mg/l.

Estimation of diquat and paraquat

Diquat and paraquat are used mixed in some commercial herbicides. They may be estimated (Yuen, Bagness and Myles, 1967) as follows:

Diquat may be determined by its absorption peak at $\lambda = 3100$ Å; at this wavelength interference from paraquat (Fig. 7.3) as well as from wetting agents, etc., is negligible.

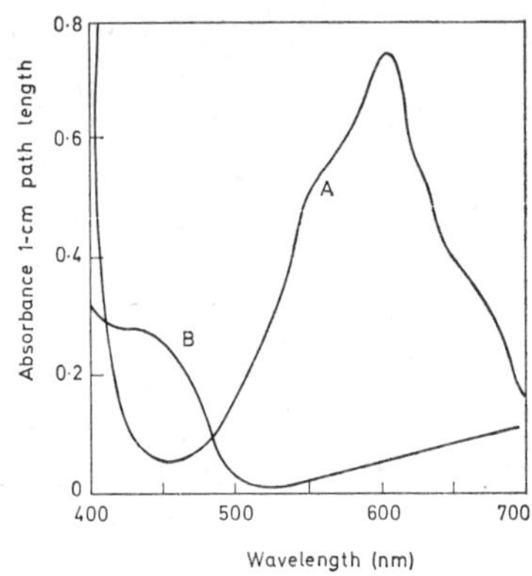

FIG. 7.4.   Absorption curves for reduced solutions.   (*A*) Paraquat, 10 mg/l.   (*B*) Diquat, 10 mg/l.   (Figs. 7.3 and 7.4 reproduced from Yuen, Bagness and Myles, 1967.)

Paraquat is not conveniently determined from its absorption peak at 2500 Å, because at this wavelength absorption by diquat and additives is not negligible. It may however be determined after reduction with sodium dithionite, which yields a blue solution (Fig. 7.4), with absorption maximum at 6000 Å. The absorption of diquat is easily corrected for since the absorption of the reduction product of the latter is linear with wavelength in this region. The base-line correction method used is in principle that of Morton and Stubbs (see page 95), but it is applied by the above authors in a slightly different manner. They measure absorbances at 5500, 6000 and 6500 Å and use a correction formula

$$A_{\text{corr }6000} = A_{6000} - \tfrac{1}{2}(A_{5500} + A_{6500})$$

$A_{\text{corr 6000}}$ is a measure of the concentration of paraquat and its value is independent of the concentration of diquat. It is not the *true* absorbance of reduced paraquat at 6000 Å, since the absorbances of the pure material at 5500 and 6500 Å are not equal (contrast the Morton and Stubbs method); but this is immaterial if a calibration graph of A_{corr} against paraquat concentration is used.

Estimation of pentachlorophenol

Pentachlorophenol, a wood preservative, is poisonous and the determination of traces in tissue is important.

(1) With fuming nitric acid the compound forms a reddish pigment which can be extracted into solution with chloroform. Deichmann and Schafer (1942) find that this solution in a normal 1 cm cell containing 1 mg gives an absorbance of $A = 0 \cdot 1$ at $\lambda = 4600$ Å.

(2) A more sensitive method makes use of the method of atomic absorption—see Yamamoto, Kumamaru and Hayashi (1967). The compound is extracted with tris-(1,10 phenanthroline)-iron (II) cation. The product is sprayed into an air-acetylene flame and the iron is estimated with a hollow-cathode iron source using the iron line at $\lambda = 2483$ Å. The calibration is linear over the range tested (0–3×10^{-4} M).

D.D.T.

D.D.T. insecticide is p,p'-dichlorodiphenyl trichloroethane. In technical products, it is often found mixed with the o,p' isomer, which is not insecticidally active. Hamilton and Beckmann (1966) show how to separate the isomers by chromatography. The p,p' isomer is dissolved in CS_2 and its infra-red absorption measured in a 1 mm cell at the absorption peak at 808 cm^{-1} ($\lambda = 12 \cdot 4$ μm). They find proportionality between absorbance and concentration after subtracting $0 \cdot 01$ from the measured absorbance.

ESTIMATION OF DRUGS

Penicillin G

This has been estimated in the presence of other penicillins by Philpotts, Thain and Twigg (1947). The presence of a phenyl group in penicillin G gives it characteristic absorption bands in the 2500–2700 Å region when used in alcohol solution. The workers mentioned have photographed spectra in this region, using a Hilger medium quartz spectrograph with a continuous spectrum hydrogen tube. A photometer was dispensed with, the spectra of an unknown solution in various thicknesses being compared with spectra, interleaved on the same photographic plate, of either a solution of pure penicillin G or another substance (pure phenyl-acetamide) showing the phenyl bands.

Fluorimetric estimation of tetracycline

Tetracycline forms a fluorescing complex with calcium barbitol. The concentration in tissue has been determined directly. This procedure avoids tedious extraction and purification and possible losses that might occur in the preparation of optically clear solutions. Winkelman and Crossman (1967), in a series of estimations, placed rat muscle homogenate on a 'reflecting' mount such as H′ of Fig. 9.12. The specimens were irradiated with $\lambda = 3780$ Å and the fluorescence observed at $\lambda = 5200$ Å. The authors found the fluorescent intensity proportional to concentration of tetracycline up to 6 μg/ml.

ESTIMATION OF VITAMINS

Morton's *The Application of Absorption Spectra to Vitamins, Hormones and Coenzymes* (1942) contains a wealth of information on the application of spectrophotometric methods. We have already noted (pages 106 and 112) some of the work on vitamins B_1 and D ; some further applications of spectrophotometry will now be discussed.

Vitamins A_1 and A_2

The spectrophotometric estimation of vitamin A has received considerable attention, and methods have been developed to overcome problems created by the fact that the vitamin exists in several forms. It is possible to make use of either the characteristic ultra-violet absorptions or the absorption band giving rise to the blue colour that can be developed with antimony chloride in the

Carr-Price reaction. Table 7.1 gives some data for each form of the vitamin—i.e. for all *trans*-vitamin A_1, *neo*-vitamin A_1 and vitamin A_2. The data are taken from papers by Morton and Cama (1953) and Morton and Bro-Rasmussen (1955).

TABLE 7.1 Absorption data for vitamins A_1 and A_2

Form of vitamin	Biological potency (Int. units per μg)	Ultra-violet absorption (*cyclo*hexane solution)		Carr-Price reaction	
		λ_{max}	$E^{1\%}_{1cm}$	λ_{max}	$E^{1\%}_{1cm}$
All *trans*-vitamin A_1	3·33	326	1745⎫	620	
Neo-vitamin A_1	2·66	328	1650⎬		
Vitamin A_2	1·33	351	1350	693	3900

The principles of the methods developed by Morton and Stubbs (1946) have already been described on page 95. The method is recommended in the *British Pharmacopeia* (1953) and the *International Pharmacopeia* (1951). Rogers (1955) gives some modified figures that he recommends as being more accurate and sets them out in table form to facilitate calculation.

Estimation of both vitamin A and carotene in butter or margarine may be made (Scott and Taylor, 1956) after chromatographic separation of these components from the non-saponifiable fraction. Vitamin A is estimated by its absorption at 324 nm, with correction for background absorption (see page 95) from measurements at 309 and 334 nm ; carotene is measured by its absorption at 446 mμ.

Vitamins B by reflection spectrophotometry

Frodyma and Lieu (1967) have shown how to separate five vitamins of the B group by chromatography on a thin layer of silica gel. After separation the invisible spots that include the vitamins are located by either:

(1) using reflection spectrophotometry to locate areas of reduced reflection at the wavelengths corresponding to absorption maxima (reflection minima—Fig. 7.5), or
(2) using silica gel to which a fluorescent dye has been added (F of Fig. 7.5), noting the areas of reduced or zero luminescence under ultra-violet illumination.

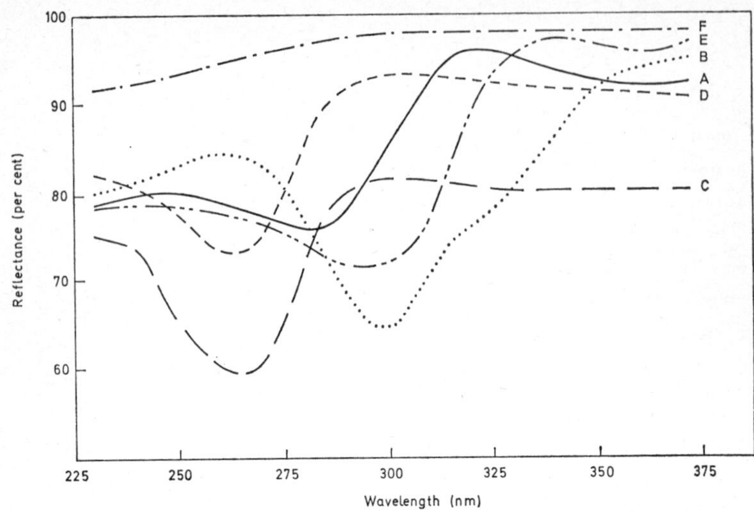

FIG. 7.5. Reflection spectra of B vitamins adsorbed on silica gel G. (A) Thiamine hydrochloride. (B) Pyridoxine hydrochloride. (C) Nicotinic acid. (D) Nicotinamide. (E) p-Aminobenzoic acid. (F) Silica gel G containing luminous pigment. (Reproduced from Frodyma and Lieu, 1967).

Portions weighing 80 mg are cut out from these areas, ground up and packed into a windowless cell for measurement of reflection at the wavelength appropriate to the vitamin in question. Calibration curves for each compound are prepared; reflection R plotted against log concentration gives curves which are convenient because they approximate to straight lines. The table below shows the sort of results obtained; techniques were developed to give reproducibility corresponding to an uncertainty of about 5 per cent in the vitamin content.

Vitamin	Wavelength used (Å)	Range of R (%)	Concentration range (μmole/80 mg)
Thiamine hydrochloride	2780	40–75	0·4–2·5
Pyridoxine hydrochloride	2980	40–80	0·2–1·2
Nicotinic acid	2650	60–80	0·7–2·5
Nicotinamide	2650	60–80	0·7–2·5
p-aminobenzoic acid	2950	40–75	0·1–0·5

Vitamin D

Morton (1942) has discussed the difficulties in the estimation of vitamin D by spectrophotometric methods. A number of workers have used methods based on the antimony trichloride reaction which gives a pink or yellowish colour with absorption maximum at $\lambda = 5000$ Å—see Chen (1936) and Nield, Russell and Zimmerli (1940). The latter workers found that common congeners of vitamin D (cholesterol, etc.) develop a similar absorption with $SbCl_3$, but more slowly. De Witt and Sullivan (1946) showed how to make use of a preliminary chromatographic separation. And for freshwater fish-liver oils, Barua and Rao (1966) first remove vitamins A_1 and A_2 by conversion to anhydro products which are separated by chromatography.

PHYSIOLOGICAL AND PATHOLOGICAL APPLICATIONS

Haemoglobin and derivatives

Haemoglobin normally occurs in blood in the reduced and oxyforms. In pathological conditions there may be other forms (e.g. met-, cyan-, carboxy- and sulph- haemoglobin, haematin, etc.). Fig. 7.6 shows the well-known absorption bands of oxyhaemoglobin and carboxyhaemoglobin. The difference between the two curves is not very great but it is nevertheless possible, by spectrophotometric measurements, to estimate small percentages of carbon monoxide in blood in cases of CO poisoning. The shift of the α-band (the peak is at 5770 Å for oxyhaemoglobin and at 5680 Å for carboxyhaemoglobin) is generally used for this estimation. A special form of instrument, the Hartridge reversion spectroscope, is sometimes used to measure the peak wavelength, which is correlated with percentage carboxyhaemoglobin. An alternative method is to measure the optical densities of a solution at two wavelengths—5770 and 5600 Å are suitable. The method is that set out on page 85 for mixtures of substances with known absorption bands overlapping ; it can be developed for routine use by means of a calibration graph showing percentage carboxyhaemoglobin against the ratio $K5770/K5600$.

Millikan (1942) has developed a two-waveband method for estimating oxyhaemoglobin and reduced haemoglobin in the presence of each other. He measures at one wavelength where the

K

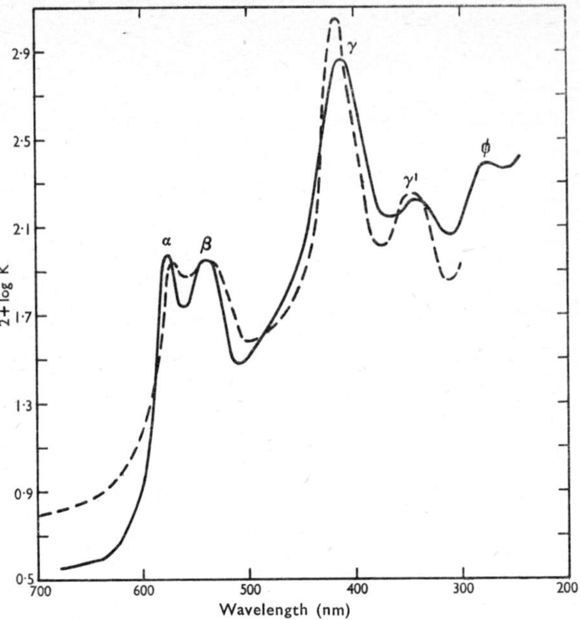

FIG. 7.6. Absorption curves for oxyhaemoglobin (——) and carboxy-haemoglobin (- - - -). (From Heilmeyer, 1943.)

two forms have the same absorption and at another where they have different absorptions. The first measurement gives total haemoglobin and the two measurements together give the proportions of the two forms. Millikan's wavebands are isolated by filters and his instrument is a small photoelectric one that clips on to the lobe of the ear to make measurements *in vivo*. This instrument has been further developed and named an ' Oximeter ' by Stott (1953), who uses filters to isolate wavebands at 780–880 nm, where the absorptions of oxy- and reduced haemoglobin are similar, and at 600–680 nm, where the absorptions differ. The interpretation of measurements with this instrument has been discussed by Kay and Coxon (1957). Verel, Saynor and Kesteven (1961) describe a similar method using another pair of wavelengths, at 4800 and 5060 Å.

Rimington (1942) has described a technique for estimating the total haemoglobin of blood by converting all the forms into one, namely pyridine-haemochromogen, and measuring a single extinction coefficient at 5500 Å. Data on the maxima of various blood pigments is given by G. A. Harrison (1957) and an account of spectrophotometric methods applied to blood and other body pigments is given by Heilmeyer (1943).

Total blood volume

The normal method of estimating the total volume of blood in a patient is to inject a small amount of suitable dye into the blood stream, and afterwards to determine the amount of dye in a small volume of blood taken from the patient. Morris (1944) first centrifuges the specimen of blood to remove the corpuscles, together with haemoglobin.

The extinction coefficient of the plasma is found for a waveband in the red where the absorption of natural blood pigment is small. If some haemolysis has occurred, as is inevitable in certain pathological conditions, it is necessary to correct for the haemoglobin that will then be in the plasma. The amount of haemoglobin may be estimated by measuring an extinction coefficient at some wavelength in the blue. Morris used filters to isolate the red and blue wavebands—Ilford Nos. 607 and 601 respectively.

Proteins

A useful and extensive review is given in *Ultra-violet Absorption Spectra of Proteins and Amino-acids* by Beaven and Holiday (1952). The use of infra-red spectra in elucidating protein structures has already been referred to (page 125).

One of the interesting applications given by Beaven and Holiday is the estimation of tyrosine and tryptophan in mixtures of the two. Measurements are made in alkaline solution at wavelengths of 2944 and 2800 Å (see Fig. 7.7) and calculation is by the method set out on page 85; 2944 Å is preferred to a longer wavelength at which the absorptions of the two components would contrast more with those at 2800 Å, because, at the wavelengths chosen, the effect of lateral shift of the absorption curves that commonly occurs with proteins will be less. The errors due to apparent absorption caused by turbidity are much reduced by measuring the optical densities at 3200–3600 Å and extrapolating

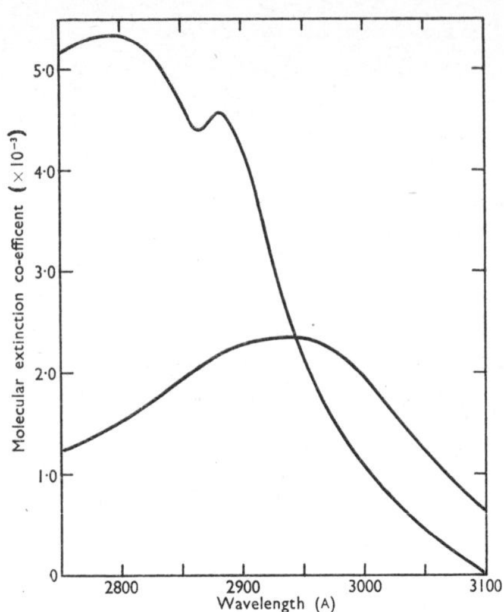

FIG. 7.7.   Molecular extinction coefficients of tyrosine and tryptophane in
0·1-N alkali.   (From Beaven and Holiday, 1952.)

to shorter wavelengths to determine the corrections to be subtracted at 2800 and 2944 Å.   The curves for pure tyrosine are
different in acid and alkaline solutions but they cross at 2768 Å ;
similarly the acid and alkaline curves for tryptophan cross at
2800 Å.   The curves for a mixture of the two substances in acid
and in alkaline solutions should therefore cross somewhere in the
region 2768–2800 Å.   If this is found to be so after applying the
correction for turbidity, it is reasonable to suppose that the
measurements are reliable ;   if the curves in acid and alkaline
solutions do not cross in this region, the correction, and hence the
whole determination, must be doubted.

Measurements on living cells

Absorption measurements on the pigments before destruction
of the cell are obviously important.   When the absorption is
sufficiently great, the absorption of a single cell may be measured
by using a microspectrophotometer of the type described in
Chapter 9 (see page 181).   Otherwise the measurements must
be made on a suspension of cells, an aqueous medium generally

being used. It is desirable to reduce loss of radiation by scatter at the surface; this can be done by reducing the refractive index change at the boundary by immersing the cells in a medium whose refractive index is as close as possible to that of the cell. Barer (1955) has shown that immersion in an aqueous egg-albumen fraction permits measurements which otherwise would be almost impossible. The protein concentration can be adjusted to give a refractive index anywhere between 1·33 and 1·42. Measurements on cell suspensions are not possible in the infra-red region where water is strongly absorbing ($\lambda=6$–7 μm approx.) unless heavy water (D_2O) is used. Blout and Lenormant (1953) found heavy water useful in detecting the presence of ribonucleic acid in certain bacilli.

Spectroturbidimetric estimation of particle size—fat globules in milk and ice-cream mixes

It was pointed out in Chapter 3 that, according to the laws of geometrical optics, the absorbance A of a suspension of uniform, perfectly opaque particles in a parallel beam of radiation is given by [equation (3.16), page 39] $A=\log_{10} e \cdot nSl$, where there are n particles per unit volume each of cross-sectional area S, and where l is the path length of the suspension. But for reasons pointed out on page 39, geometrical optics does not tell the whole story and a much more generally useful equation is

$$A=\log_{10} e \cdot n\underset{\sim}{Q}Sl \qquad (7.1)$$

where the cross-sectional area S has to be multiplied by a factor Q, known as the total scattering coefficient, the scattering area coefficient or (Van de Hulst, 1957) extinction efficiency factor. Q may have values between 0 and about 4, depending on the absorbing properties of the material of the particle, its refractive index relative to that of the surrounding medium, its size relative to the wavelength of the radiation and on its shape. The numerical values of Q can be calculated in certain cases and Fig. 7.8 shows the results of such calculations for transparent spherical particles where the relative refractive index $\mu=1\cdot09$. These conditions apply to globules of milk fat in diluted milk ($\mu=1\cdot45/1\cdot33=1\cdot09$), and Goulden (1958) has made use of the variation of Q with particle radius r and wavelength λ to estimate the size and size distribution in various samples of raw and processed milk. This is done by comparing the shape of an experimental absorbance/wavelength curve with a set of curves such as those of Fig. 7.9, which are calculated from Fig. 7.8. It is seen that measurements in the range $\lambda = 3500$ Å to $1 \cdot 35$ μm give curves with considerable shape variations for the range of globule diameters (about 1–4 μm) found in milk.

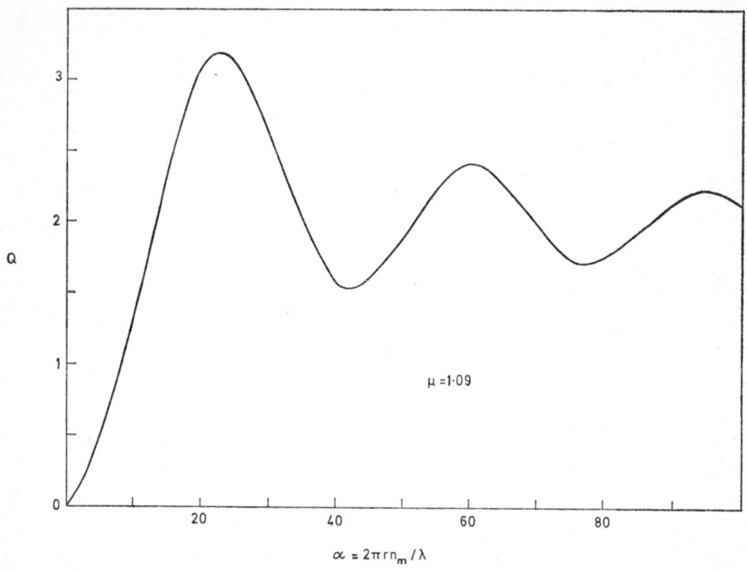

FIG. 7.8. Calculated values of scattering area coefficient Q for various values of the parameter $a = 2\pi r n_m/\lambda$, where $r =$ particle radius, λ is the wavelength of the radiation in air and $n_m =$ absolute refractive index of the medium in which the particles are suspended (in the example discussed in the text $n_m = 1 \cdot 33$). (Reproduced from Goulden (1958)).

The calculations on which Figs. 7.8 and 7.9 are based assume a perfectly parallel beam of incident radiation and a detector system which receives only the radiation directly through the suspension and excludes radiation scattered away from this direction through even a small angle. The condition is sufficiently fulfilled by standard commercial spectrophotometers after minor modifications described by Goulden (1958), Goulden and Sherman (1962) and Walstra (1965).

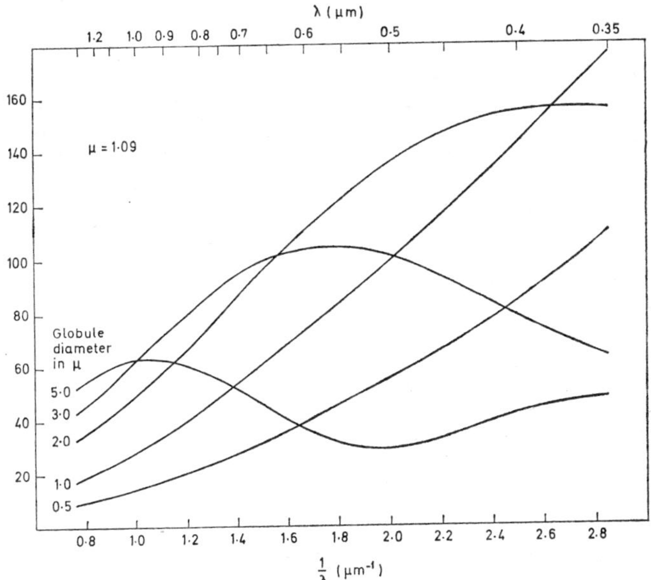

FIG. 7.9.   Variation of absorbance with wavelength for suspensions of fat globules of various uniform sizes.   The curves are calculated from Fig. 7.8 and the ordinates are in arbitrary units proportional to absorbance (i.e. proportional to Q).   (Reproduced from Goulden, 1958.)

It may be mentioned that Goulden and many other authors on this subject use the symbol K in place of the symbol Q used here (after Van de Hulst). K has been avoided here because it is used in this volume with quite a different meaning (see Table B).

PART III

The Technique of Spectrophotometry

CHAPTER 8

The General Design of a Spectrophotometer

SOME general features of spectrophotometer design will be discussed under five headings:

1. Sources of radiation.
2. The monochromator.
3. Photometric devices.
4. Radiation detectors.
5. The mounting of specimens and the method of passing radiation through them.

1. SOURCES OF RADIATION

Tungsten-filament lamp. A gas-filled coiled-filament lamp provides a source of high intensity per unit area. A motor head-lamp with straight filament gives a source that is convenient because it is readily imaged on the entrance slit of a monochromator. With its normal glass envelope it is useful in the wave-length range 3500 Å–2 μm. The range can be extended to nearly 4 μm by use of a silica envelope or window; and to 8 μm if a periclase window (MgO) is substituted.

Increased output may be obtained by running the lamps at 10 per cent or more above their normal voltage. This increases the colour temperature and so considerably increases the radiation of short wavelengths.

Quartz-iodine lamp. This is a special design of tungsten filament lamp, in which the filament is inside a fused silica envelope containing iodine vapour. With such a lamp the tungsten filament may be run at a higher temperature than usual—any tungsten that is distilled on to the inside of the envelope is removed again by the iodine vapour. Such a lamp gives appreciably more ultra-violet than a normal gas-filled tungsten lamp.

Deuterium discharge lamp. This discharge with a hot cathode runs from a supply of 200 volts. It is compact, with the radiation emitted from a slot in the cylindrical anode, so that the source is conveniently imaged on a monochromator slit. The molecule D_2 emits a continuous spectrum covering the range 1500–5000 Å. Superposed on this are some spectrum lines of D_2 as well as the atomic (Balmer) lines; these last are useful for wavelength calibration. The deuterium lamps are found to have a longer life and somewhat higher intensity than those using normal hydrogen.

Xenon discharge lamp. A discharge through high-pressure xenon gives a source of visible and near ultra-violet radiation with a very high intrinsic brightness up to 40 000 candelas per sq. cm. The radiation has a continuous spectrum and is useful as a source for spectrofluorimeters.

Rare-gas discharge lamps. Discharges through the rare gases give continuous spectra arising from unstable molecules, He_2, etc. covering the following wavelength ranges:

Helium	600–1000 Å
Argon	1150–1500 Å
Krypton	1250–1650 Å
Xenon	1450–1800 Å

The short range of wavelengths emitted by each eliminates the need, when diffraction gratings are used, to remove overlapping orders. The discharges also include some spectral lines which are useful for checking wavelength calibrations. The use of these lamps with r.f. excitation has been described by Wilkinson and Byram (1965).

Lyman continuum. The Lyman continuum (Lyman, 1926) extends over the range 300–7000 Å. It is obtained by passing a discharge of high current density, obtained by periodic discharge of a condenser, through a narrow-bore tube of glass or silica; it probably results from thermal radiation from particles torn from the tube by the discharge. Garton (1953) has described a convenient demountable design for obtaining the Lyman continuum.

High-pressure mercury lamp. For use in the ultra-violet. Such a lamp, obtainable commercially (85 or 125 watts) with the discharge in a silica tube and surrounded by a glass outer envelope gives the intense spectrum lines at 3650, 4047, 4358, 5461, 5770,

5790 Å. These are somewhat broadened at the pressure of an atmosphere or so and there is a certain amount of continuous spectrum between the lines. If the lamps are used without an outer envelope or with a silica envelope, further lines and a continuum are obtained down to about 2000 Å. These lamps are useful for spectrofluorimetry.

For the far infra-red. A very-high-pressure mercury lamp in a silica envelope and often water-cooled is almost universally used for the infra-red for wavelengths longer than about 40 μm. The emission is in part from the mercury vapour plasma and in part from the silica envelope.

The Nernst filament. This is a semi-conductor made of a mixture of zirconium oxide and rare-earth oxides. It is non-conducting when cold, but when heated to just-red heat by a bunsen flame or an electric heater it becomes conducting and can then be maintained incandescent by an electric current. The spectral-energy distribution curve has been determined (Ritzow, 1934) and plotted to show the emissivity relative to that of a black body at the same temperature. The results indicate that emissivity is high in the visible, relatively low (30 per cent) in the 1–6 μm region, and rises again to 90 per cent at 8 μm. The Nernst filament is obtainable in a size suitable for projecting on a spectroscope slit, and also for running from the mains (preferably a.c.) with a stabilizing resistance.

The Globar. The Globar is a carborundum rod used for electric furnace heaters and sometimes for domestic electric fires. It has the advantage of being conducting when cold and so needs no special starting circuit. It is at present obtainable only in the large diameters suitable for heaters and therefore needs to be used in a water-cooled housing to remove the fairly large amount of heat dissipated. Some workers claim that it gives a greater energy output in the longer wavelengths (beyond about 10 μm) than a Nernst filament.

2. THE MONOCHROMATOR

To obtain monochromatic radiation a prism and/or grating may be employed. Filters may also be used either alone or to supplement a prism or grating.

Prism instruments

The transmission and dispersing properties of prism materials have already been discussed; see pages 64–7 and Figs. 4.1–4.5. The most simple design consists of one or more prisms through which the radiation passes at approximately minimum deviation. This is used in a spectrograph where the complete spectrum is observed, or more commonly photographed, at one time. Where, as in non-photographic instruments, the spectrum is observed wavelength by wavelength, it is desirable to have a system in which the spectrum is passed across a fixed position. This may be achieved by using a constant-deviation arrangement similar to those now to be described.

Littrow mounting. (See Figs. 9.2 and 9.4.) Radiation passes through the prism at almost minimum deviation and is then reflected back through it to undergo further dispersion. The spectrum may be traversed over a fixed exit slit by rotation of the prism (Fig. 9.2) or of a separate mirror (Fig. 9.4). If a 30-degree prism is used, it is common practice to make the rear surface

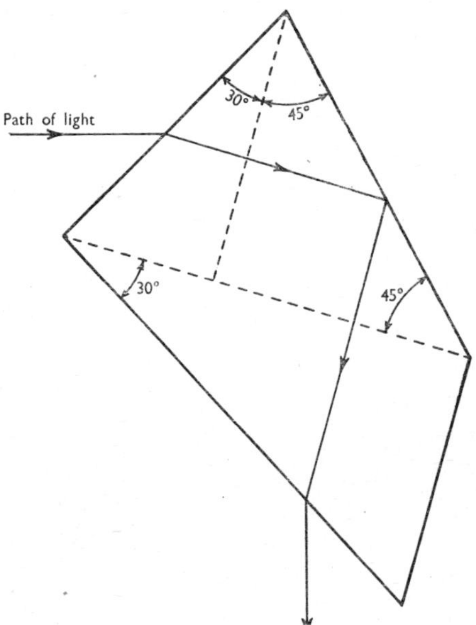

FIG. 8.1. Constant deviation prism.

highly reflecting by coating it with a metal such as aluminium, which has a high reflection in the ultra-violet, visible and infrared. But alkali halide prisms, which are hygroscopic, are not satisfactory when so treated, and a separate reflecting surface is usual. This is in any case necessary when a 60-degree prism is used (see Fig. 9.4). Advantages of the Littrow mounting are economy in cost of the prism material, obtaining the dispersion of a 60-degree prism with a 30-degree prism, and the economy of space that results from bringing the entrance and exit slits close together. As in the simple spectrograph design, several prisms may be used in series to obtain still greater dispersion.

Wadsworth mounting. An alternative constant-deviation device consists of a 60-degree prism followed by a mirror; if prism and mirror are rotated as a unit, the spectrum traverses a fixed exit slit (see the standard textbooks). This design has been used in several commercial instruments but Littrow mountings are now more common. There is however a particular Wadsworth design in which the reflection is achieved by total internal reflection at a surface of a special prism (Fig. 8.1) that consists in principle of two 30-degree prisms on each side of a totally reflecting prism. This is often used with glass but not with other more costly materials.

Focusing

The focusing of the spectrum may be achieved by means of lenses or mirrors. Lenses are generally used for the visible spectrum, where they may be made approximately achromatic, and for spectrographs, where the photographic plate may be so placed that all parts of the spectrum are correctly focused on it. Mirror focusing is much more general because the system is then completely achromatic and adjustment for one wavelength means that the whole spectrum will be in focus. The radiation, however, cannot be incident on the mirror quite normally and this may give rise to serious aberrations. The difficulty may be overcome in several ways:

1. By using a 'figured' instead of a spherical mirror to reduce the aberrations to reasonable amounts.
2. By placing the slit above or below the prism so that the imperfections of the slit image lie largely along the length of the

image and do not affect the spectroscopic performance. This method is used in the design of Fig. 9.2 on page 178.

3. By arranging, as in a design by Pfund (1927), for radiation from the slit to fall on the concave mirror and be reflected back to a plane mirror surrounding the slit, which then reflects it, without introducing aberration, to the prism or grating. This design is sometimes used in laboratory-made instruments because it avoids the difficulty of obtaining specially figured mirrors.

Diffraction gratings

The mounting of a plane diffraction grating is similar to that of a prism and is to be seen in Fig. 9.3.

A reflecting concave grating needs no additional focusing component and hence is particularly useful in the short-wave ultra-violet, for which there are few adequate lens materials available and where it is desirable to keep the number of reflecting surfaces to a minimum in order to retain adequate intensity. The grating and two slits, or, for a spectrograph, one slit and a photographic plate, must all lie on the Rowland circle, which has a diameter equal to the radius of curvature of the grating, as explained in standard textbooks. Figs. 9.6 and 9.7 on pages 182 and 184 illustrate the use of a concave grating at near-normal incidence.

Echelette gratings. When radiation is diffracted from a grating it is distributed among several 'orders' of spectra. Modern gratings have the rulings so formed that up to 40 per cent of the incident radiation falls into one order. Such rulings, first prepared by R. W. Wood (1910 and 1937), are known as echelette or 'blazed' gratings.

Separation of orders. In general, several different wavelengths, λ_1, λ_2 etc., belonging to different orders of spectra, will be superposed. These are related by the equation

$$n_1\lambda_1 = n_2\lambda_2 = \ldots$$

where n_1, n_2 etc. are integers, usually small, whose values denote the orders of the spectra. Thus, if one is observing a wavelength of 6000 Å in the first order, superposed on this will be wavelengths of 3000 Å and 2000 Å belonging to the second and third orders respectively. But if the eye is being used for observation, the latter will not be detected. Again, even if the detector were a photocell sensitive to these wavelengths, they would be absorbed if the system included some glass. In the infra-red, however, where thermal detectors are sensitive to all wavelengths, overlapping orders must be removed; this may sometimes be achieved by means of suitable filters (see page 190) but more generally a prism monochromator is used in series with the grating monochromator, as in the instrument illustrated in Fig. 9.3.

3. PHOTOMETRIC DEVICES

Different methods of measuring and comparing the absorption of specimens have resulted in the development of three main classes of photometric instruments.

Class 1

The instrument has two beams, in one of which is the absorbing specimen and in the other a photometric device—polarizing prisms, adjustable aperture, etc.—for matching the intensities of the two beams. Visual spectrophotometers always use this principle and it is used in some photographic instruments (e.g. the Spekker ultra-violet photometer).

When an instrument of this type is used with a photoelectric or thermal detector, the two beams are flickered alternately on a single detector in such a way as to obtain, at the match point, a steady photo-current without any ripple. This will be achieved if the changeover from one beam to another is suitably made— either by (*a*) cutting one beam off suddenly at the exact instant the other is admitted or (*b*) reducing one beam gradually at the same rate as the other is increasing, so that, at the match point, the photo-current remains steady. Although neither of these conditions can be fully obtained, the current fluctuation can be reduced to a minimum value at the match point.

Class 1*a*

In a modification of this principle there may be only one beam and one detector, the absorbing specimen and photometric device

L

being used in succession in the beam. This is used in the Spekker absorptiometer (see page 195) and in some photographic spectro-photometry where successive exposures are made with the absorbing specimen and a series of known absorbances in the beam.

Class 2

In the second type a single beam and single detector are used, the transmission of a specimen being the ratio of the detector currents with and without the absorbing specimen. This is in essence the simplest possible design ; it assumes, however, that the response of the detector is proportional to the intensity falling on it. It also requires the intensity of the light source to remain steady during alternations of the absorbing specimen and compensating cell. Nowadays, however, it is possible to fulfil both of these conditions reasonably well—the greater the accuracy required the greater the attention that must be given to these points.

4. RADIATION DETECTORS

Detectors used in spectrophotometers are discussed below under four headings according to their mode of action. Their quantitative characteristics are described by several parameters which will first be defined.

Responsivity is the response (usually in volts or amperes) for unit incident radiant power (expressed in watts or, in the visible spectrum, sometimes in lumens). For detectors that are wavelength selective, it is of course necessary to state the wavelength, for which the responsivity is quoted. The term 'sensitivity' has sometimes been used in place of responsivity, but the usage is to be deprecated because a detector with high responsivity may not be 'sensitive' for the detection of low intensities, this latter being largely determined by noise considerations (see page 72).

Noise equivalent power (NEP) is the minimum incident radiant power that gives a signal-to-noise ratio s/n equal to 1, calculated for an amplifier band width of 1 c/s. Since [page 72, equation (4.9)] the noise is proportional to the square root of the bandwidth, the dimensions of NEP are watt . $(c/s)^{-\frac{1}{2}}$.

Noise equivalent input (NEI), sometimes called equivalent noise input (ENI), is the incident power per unit area (S) of the detector, i.e.

$$NEI = NEP/S \qquad (8.1)$$

Detectivity (D) is

$$D = 1/NEP \qquad (8.2)$$

so that its dimensions are $(c/s)^{\frac{1}{2}} \cdot watts^{-1}$. Thus a detector with a high value of D is better for use at low intensities.

On the assumption that the noise of a detector is proportional to the square root of its area (see under 'Thermal Noise', equation (4.9a), page 73), a figure of merit often quoted and usually named 'dee-star' (D^*) is defined as

$$D^* = S^{\frac{1}{2}}/NEP \qquad (8.3)$$

and the units are cm. $(c/s)^{\frac{1}{2}}/watt$.

Values of NEP, NEI, D and D^* must state fully the conditions to which they apply; viz. (1) the wavelength, except for thermal receivers that are black; (2) receiver temperature; and (3) chopping frequency. Thus for a Ge/Cu photoconductive cell, the statement $NEP(15, 800, 1) = 2 \cdot 4 \cdot 10^{-11}$ watt $\cdot (c/s)^{-\frac{1}{2}}$ means that at $\lambda = 15\ \mu m$, with the radiation chopped at 800 c/s and an amplifier band width of 1 c/s the signal/noise ratio is unity for an input radiant power of $2 \cdot 4 \cdot 10^{-11}$ watt.

Table 8.1 gives some of these parameters for the detectors discussed in the following sections.

DETECTORS

THE detectors used in photoelectric spectrophotometers may be divided into four main classes:

1. Photo-emissive cells.
2. Photo-voltaic cells.
3. Photoconductive cells.
4. Thermal detectors.

The uses and limitations of each type will now be briefly discussed.

Photo-emissive cells

These consist of a photo-sensitive surface that emits electrons when light of suitable wavelength falls on it. The photo-sensitive surface, or cathode, is at a negative potential relative to the collecting electrode, or anode. The space between anode and cathode is evacuated.

A number of cathode surfaces that are sensitive to various wavelengths are available. The more important are :

Cathode	Approximate useful wavelength range
Sodium	3000–5000 Å
Potassium	4000–5000 Å
Caesium on caesium oxide on silver	2000–12,000 Å
Potassium on silver	ultra-violet–7000 Å
Antimony-caesium	ultra-violet–6600 Å
Bismuth-oxygen-silver-caesium	ultra-violet–7500 Å
Platinum	– 2500 Å
Nickel	– 2500 Å
Tungsten	– 1900 Å

The antimony-caesium cathode has, within its useful range, a much higher sensitivity than other types. Cells that are to be used for the ultra-violet may have a quartz window or be enclosed entirely in a fused silica bulb.

For work below 2000 Å it is common practice to use a normal blue-sensitive photocell whose window is coated with a fluorescing material that effectively converts the short-wave radiation into near ultra-violet and violet. Johnson, Watanabe and Tousey (1951) have tried a number of fluorescent substances and find that a coating of sodium salicylate is most satisfactory ; it has a high and nearly constant quantum efficiency for conversion of wavelengths in the 900–2300 Å range, but is not very sensitive to the longer ultra-violet, so that the effect of scattered ultra-violet radiation is reduced.

For very short wavelengths, Hinteregger and Watanabe (1953) have used photocells with cathodes of platinum, nickel and tungsten and Gabriel, Swain and Waller (1965) have used gold. These have the advantage of not being sensitive to longer wavelengths. Since these cathode materials are chemically stable, they may be used inside the vacuum of the spectrometer without a window.

The cells that are sensitive to the infra-red also have thermionic emission and the inevitable fluctuations of this set a limit to the smallest detectable photoelectric current. (A possible method of reducing this thermionic effect is to cool the cell with solid carbon dioxide or liquid air.)

A photo-emissive cell gives a current that may be passed through a high resistance (ranging from 10^6 to 10^{11} ohms), thus setting up a potential difference which lends itself to amplification. For detecting very small currents (down to 10^{-12} amp or less) special electrometer valves with very low grid current are used. Alternating current has some advantages for amplification and so it is not uncommon to chop the light beam to obtain a photo-current with an a.c. component. Tuning an amplifier to the frequency of chopping will minimize the effects of ' noise '.

In many spectrophotometers the photo-current is used to indicate the balance point and no measurement of current or voltage is required. In spectrophotometers of Class 2 (page 162), however, the output current is measured sometimes on a suitable galvanometer and sometimes (see Fig. 8.2) by means of a potentiometer in the grid circuit of the valve, the potentiometer being adjusted to give a fixed reading on the output meter. The latter arrangement does not require the amplification to be linear.

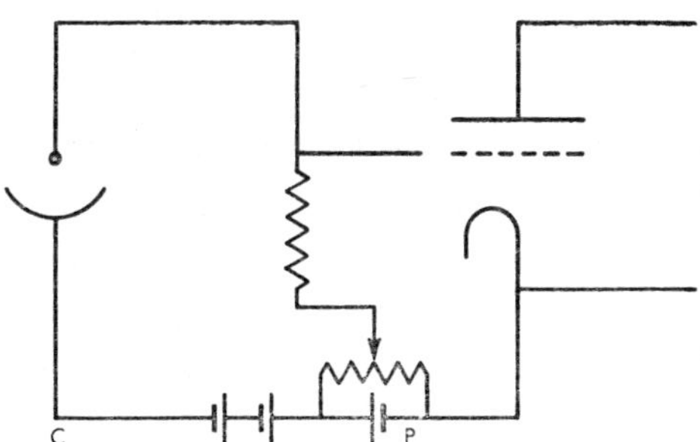

FIG. 8.2. Simplified circuit diagram showing use of a potentiometer (P) in first stage of an amplifier for measuring current from a photocell (C).

Photomultipliers

The noise level of a photomultiplier is much less than is obtained with a normal photo-emissive cell coupled to a valve amplifier, and the former is commonly used where the radiation intensity is low, as in vacuum ultra-violet spectrophotometers and in spectro-fluorimeters.

Photo-voltaic cells

These are also known as rectifier or barrier-layer cells. Commercial versions consist of a layer of selenium (or less commonly cuprous oxide) between a metal plate that forms one contact and a transparent metal film that forms the other contact. When light is incident on the 'barrier' layer between selenium and metal film, a potential difference, which may be several volts, is set up and no external battery is required. It is usual to feed current directly into a galvanometer. Selenium barrier-layer cells are usually sensitive from about 3000 Å to 8000 Å.

Photoconductive cells

Cells of this type change their electrical resistance when radiation of appropriate wavelength falls on them. This resistance change is measured as the voltage change across the load resistance in a circuit which also incorporates a d.c. voltage supply.

Since the cell resistance also changes with temperature, it is common practice to chop the radiation and to amplify the intermittent current with a tuned a.c. amplifier. In this way changes of current due to changes of ambient temperature, and random changes due to 'noise', are not amplified. In most commercial cells the optimum signal-to-noise ratio is obtained with a chopping frequency of 800–1000 cycles per second.

Commonly used photoconductive cells are lead sulphide, lead selenide and indium antimonide. All these materials are used in an effectively pure form and act as 'intrinsic' photoconductors. The action of the cells depends on the relatively small energy gap E between the valence band and the conduction band. The minimum energy of a photon that will cause an electron to jump into the conduction band is equal to the energy gap E; thus the long-wave limit is determined by $E = hc/\lambda_{max}$.

'Extrinsic' photoconductors depend on materials which are doped with impurity atoms (e.g. germanium doped with copper or gold, and doped indium antimonide). These cells are sensitive to much longer wavelengths than intrinsic photoconductors—doped InSb cells may be used at least as far as $\lambda = 1000\,\mu\text{m}$. These cells must be cooled with liquid helium; but this is now fairly easily obtainable and its use should not be considered too exotic.

It will be noted from Table 8.1 that the NEP is not smaller for photoconductive cells than for say thermopiles. But the faster response of the former permits higher chopping frequencies for which it is possible to use a much more sharply tuned amplifier; in practice, therefore, photoconductive cells at suitable wavelengths offer the possibility of using narrower slits than can be used with thermopiles.

Limitations of photocells and their measuring circuits

The performance of photoelectric spectrophotometers may be limited by several factors that affect the operation of the cells and their accompanying galvanometers, amplifiers, etc. The first of these factors is the linearity of the relation between photoelectric current and light intensity. In vacuum photocells it is unusual for the cells to depart from linearity by more than about one per cent and, by choosing a cell with suitable electrode construction, linearity to at least 0·1 per cent can be attained (see Preston and McDermott, 1934). The response of selenium photo-voltaic cells depends on both light intensity and the resistance in series with the cell. The current is generally proportional to light intensities for small intensities and small resistances. The most sensitive galvanometers, however, must have high resistance, and this tends to affect the response of the cell. Care is therefore required in designing a combination of cell and galvanometer that will give linear response.

The second factor is linearity of the measuring circuit. If a spectrophotometer depends on linearity of response (Class 2 above) it is obviously important that the measuring circuit as well as the photocell should be linear. Care is required in designing a valve amplifier that will be linear in response. Potentiometers and galvanometers are used as measuring devices; it is not difficult to procure these with the requisite linearity but their performance should nevertheless be tested.

Thermal detectors

Thermal detectors depend on the rise of temperature that occurs when energy is absorbed and on the measurement of some temperature dependent property. In thermopiles this property is the development of a thermoelectric e.m.f.; in bolometers, a change of electrical resistance; and in the Golay pneumatic detector, a change in gas pressure.

Thermopiles

A thermopile consists of a number of thermojunctions in series; the hot junctions are attached to thin blackened receivers on which the radiation is incident, and the cold junctions are in contact with a relatively massive heat sink. The most commonly used thermopile is the Schwarz type where the thermoelectric materials are a pair of semiconductors chosen to have high thermoelectric efficiency and to have low thermal conductivity so that maximum temperature rise is attained. The receivers are very thin gold or platinum leaf blackened on the front face. Performance is improved if the receiver area is kept small. This has the effects of (1) reducing the thermal noise [equation (4.9a), on page 73] and (2) increasing the responsivity, because a smaller surface means that heat losses by radiation are reduced and that the thermal capacity is less. Note the areas quoted in Table 8.1; a reduced image of the exit slit of the monochromator is formed on the receiver by a concave mirror (see Fig. 9.4, page 180). Responsivity may also be increased by using the thermopile in vacuum, so that there are no heat losses by convection or by conduction through the gas.

Bolometers

Bolometers, which utilize the change of resistance with change of temperature, have been made in the past of nickel, the metal giving the greatest temperature coefficient of resistance, but certain semiconducting ' thermistor ' materials, usually mixtures of oxides of manganese, nickel and cobalt, are now also used, since they have high temperature coefficients of resistance. They may be used in a thickness of about 10 μm mounted on a backing of glass or silica to give a response in 0·003–0·005 second. Such bolometers have a resistance of a few megohms. Wormser (1953) finds that they have a noise level of only one and a half times the theoretical Johnson noise due to Brownian fluctuation (see below). Gillham (1956) has described the construction of semi-conducting bolometer films of antimony. A bolometer consisting of a thin evaporated layer of gold is described by Archbold (1957); it has a time constant of 0·005–0·007 second and the minimum detectable energy (limited by Johnson noise) is $3·6 \times 10^{-10}$ watts.

The resistance change in a bolometer is measured as the change in voltage produced somewhere in a circuit—which may be a Wheatstone bridge circuit—that incorporates a source of d.c. voltage. The value of this voltage naturally affects the responsivity in volts/watt. But there comes a stage at which further increase in this applied voltage causes no further increase in responsivity; the theory is discussed in detail by Smith, Jones and Chasmar (1957). The responsivities quoted in Table 8.1 are for this optimum voltage.

TABLE 8.1. Properties of some radiation detectors
The figures are collected from various sources and should be taken to indicate orders of magnitude only. For selective detectors the figures in the last two columns are for the wavelength of maximum response.

Detector	Area (mm^2)	Temperature (°K)	Long wavelength limit (μm)	Resistance (ohms)	Response time (m sec)	Chopping frequency (c/s)	Responsivity (volts/watt)	Noise equivalent Power (Watts (c/s)$^{-\frac{1}{2}}$)
Photomultiplier		300	0·7		$< 10^{-5}$			10^{-14}
Photo-emissive Cells:								
Cs/Ag$_2$O/Ag			1·2					
Sb/Cs			0·65					
Bi/Ag/Cs			0·75					
Pt			0·25					
W			0·19					
Photoconductive Cells:								
PbS	10×1	300	2·9	10^5	0·04	800	5×10^3	2×10^{-11}
		196	3·3	10^6	0·2	800	15×10^4	4×10^{-12}
PbSe	6×1	300	4·0	10^5	0·02	800		$2·6 \times 10^{-10}$
PbTe	10×1	196	5·6	10^7	0·01	800		$1·1 \times 10^{-10}$
InSb	$5 \times 0·5$	300	7·9	100	$< 10^{-4}$	800	1	10^{-9}
Ge/Cu	6×1	4·2	25	10^5	10^{-3}	800	1000	$2·4 \times 10^{-11}$
Thermal detectors								
Thermopile (Schwarz, vacuum)	$4 \times 0·2$	300		35	35	10	30	$2·5 \times 10^{-11}$
Bolometer (thermistor)	$2·5 \times 0·2$	300		3×10^6	3·5		700	3×10^{-10}
Golay detector	3 (dia.)	300			20	12·5		$1·6 \times 10^{-10}$

The Golay detector

The Golay pneumatic detector is illustrated in Plate 1B [see Golay (1947)]. Radiation to be measured falls on a small and thin blackened film B of small heat capacity. The rise in temperature of the film heats the gas in the chamber containing the film and so increases the gas pressure. The chamber is sealed at one end by the entrance window A and at the rear by a mirror membrane C, so that the increase of pressure distorts the mirror and causes it to reflect varying intensities of light from a separate lamp E via a mirror into a photocell D. The detector is used with a chopped beam of radiation and the amplified a.c. component of photoelectric current is therefore a measure of the radiation intensity. The response time is short enough to permit chopping at ten cycles per second and the detectivity is comparable with that of other thermal detectors.

5. THE MOUNTING OF SPECIMENS

A specimen should ideally be placed in a parallel beam so that, as explained on page 71, the path length through it is definite. In practice no beam can be perfectly parallel if it is to carry any energy, and when a long path is necessary, as often with gases, it is more usual to use a definitely non-parallel beam with an image of the source near the centre of the cell (see Fig. 9.4 on page 180); the diameter of the cell, and hence the volume of the specimen, can then be kept to a minimum. This form of beam is also used with a diffusing specimen, which is placed at the intermediate focus of the source so that at least some radiation scattered from it may be collected for measurement. Paraffin mulls and alkali halide disks (see below) are so mounted.

Measurement on powders

The absorption spectra of solids that cannot be used in a homogeneous form as a single crystal are sometimes determined by measurement of the powder in reflected radiation ; Rosenwasser and Dreyfus (1956) have in this way measured the absorption bands formed in sodium azide by irradiation with gamma rays. Quantitative interpretation of such measurements is not simple, however, as pointed out on page 40. An alternative and much-used method has been to mix the finely ground specimen into a mull with a transparent liquid of not too different refractive index, and to use it in a normal cell as for liquids. While this method has been widely used in the infra-red, with some form of paraffin for mixing, it suffers from the disadvantage that all the liquids available have some absorption bands of their own.

The pressed-disk technique

This method, first described by Stimson and O'Donnell (1952), offers a great improvement, and consists of grinding together a mixture of an alkali halide and a specimen for measurement, and pressing the mixture into a glass-like disk in a special press ; a disk of alkali halide alone is made for comparison. Alkali halides are particularly suitable because they are transparent throughout a large part of the ultra-violet and infra-red, and because they readily flow under a pressure of about 50 tons per square inch. Potassium bromide is the most commonly used because it is less hygroscopic than some alkali halides and flows under a smaller pressure (about 40 tons per square inch). A number of presses suitable for making disks up to about $\frac{1}{2}$-inch diameter have been described—see, for example, Ford and Wilkinson (1954). To prepare good disks, the material should be dried and pressed under vacuum to avoid occluded air bubbles. Disks are most suitably mounted at an intermediate focus of the source, which is then focused on the slit of the spectrometer ; the instrument shown in Fig. 9.4 is suitably arranged. The precautions necessary in interpreting measurements on powdered specimens are discussed on page 38.

This method must be used with care ; it has been found (see Farmer, 1955) that specimens, especially those with OH groups, may be adsorbed on the surface of the alkali halide and modify the absorption spectrum. Baker (1957) also finds differences between spectra obtained by the disk and mull techniques that may be attributed to various causes, particularly polymorphism of the specimen ; he suggests that the two techniques are to be regarded as complementary.

Measurement by reflected radiation

There is no particular difficulty in measuring specular reflection ;

it is necessary only to arrange the components so that the specularly reflected radiation is directed to the detector.

When measuring on matt surfaces, however, the radiation that is not absorbed is diffused over a hemisphere and to obtain adequate intensity in the measured beam it is necessary to collect a fair proportion of this radiation on the photocell or other detector. Various optical arrangements may be used for doing this. Fig. 8.3a shows an arrangement in which radiation leaving the specimen surface at angles between 35 and 55 degrees strikes a ring-shaped ellipsoidal mirror that focuses it on a diffusing screen immediately in front of the measuring photocell. In a somewhat similar arrangement some of the reflected radiation is collected by an annular mirror (Fig. 8.3b) and passed directly to the photocell.

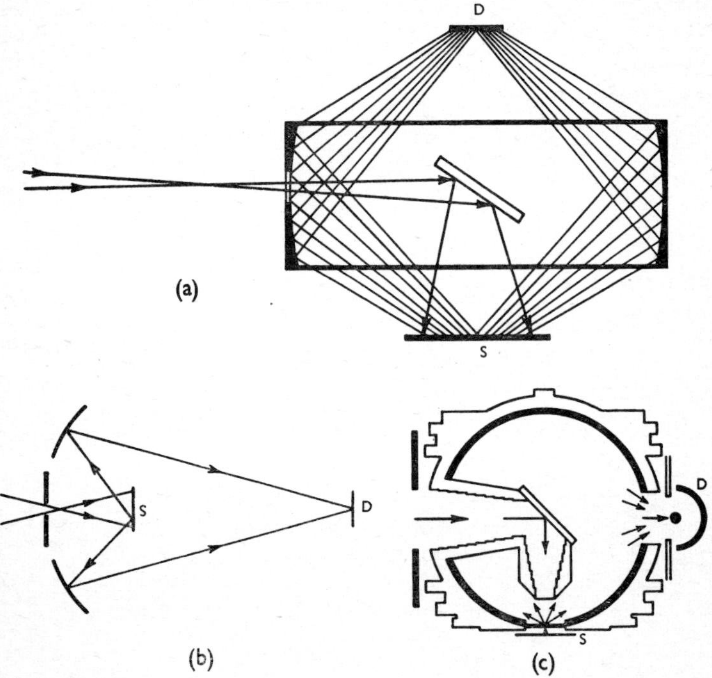

FIG. 8.3. Mounts for the measurement of diffuse reflection as used (a) on the Beckman DU spectrophotometer, (b) on the Uvispek spectrophotometer, and (c) on the Beckman B spectrophotometer. In each figure, S is the specimen and D the detector.

A different arrangement places the specimen over an aperture in the surface of a diffusing or integrating sphere—see Fig. 8.3c. Radiation reflected from the specimen falls on the inner surface of the sphere and some of this ultimately makes its way, after multiple reflections, to the detector in another aperture of the surface. Measurements with an integrating sphere must be carefully interpreted because some of the radiation reaching the detector has been reflected from the specimen more than once and so the instrument reading and the reflection of the specimen are not strictly proportional. The theory is as follows:

Let the reflection of the specimen be r_s and that of the sphere wall r_w, and let the surface areas of the sphere wall, specimen aperture, entrance aperture and detector aperture be a_w, a_s, a_e and a_d respectively. Let the intensity of the beam entering the sphere and falling on the specimen be i and let the intensity incident on unit area within the sphere be I. Then I is determined by the equilibrium between the radiation entering the system and that leaving it by ' absorption ' at the walls and apertures,

i.e.
$$r_s i = I[(1 - r_w)a_w + (1 - r_s)a_s + a_e + a_d]$$

or
$$I = \frac{r_s i}{(1 - r_w)a_w + (1 - r_s)a_s + a_e + a_d} \qquad (8.4)$$

From this two conclusions may be drawn :

1. For high efficiency (large I) the terms in the denominator should be kept small, i.e. the reflection of the sphere should be high and its area should be as small as is consistent with the space required for the necessary apertures.

2. The measured intensity I and specimen reflection r_s are only proportional if the term $(i - r_s)a_s$ is negligible compared with the other terms in the denominator. It should be noted that this requires that a_w should be large, contrary to the requirements of (1) above.

There is no important lack of linearity when comparing specimens with similar reflections. With specimens that are not similar, proportionality between instrument response and reflection may be retained by using a modified design in which two specimens (1 and 2) to be compared are exposed simultaneously in adjacent apertures of the sphere surface ; the beam is made to fall on each in turn either by moving the beam or by interchanging the specimens. Then expression (8.4) has terms $(1-r_{s1})a_s-(1+r_{s2})a_s$ in the denominator for both the measurements, so that $I_1/I_2=r_{s1}/r_{s2}$. Arrangements of this type are discussed by Hardy and Pineo (1931).

Microspectrophotometry

For measurements on very small specimens, such as living cells, it is necessary to use a microscope in conjunction with the spectrometer. In order properly to define the portion of specimen being observed, focusing must be good for all wavelengths and a reflecting microscope is the obvious choice. A number of suitable types have been described—see, for example, Burch (1947), Wilkins and Norris (1952), Fraser (1953) and Thornberg (1955).

Various combinations of microscope and spectrometer are described in the report of a discussion held by The Faraday Society on 'Optical Methods of Investigating Cell Structure' (Faraday Society Discussions, 9, 1950). If a photographic instrument (spectrograph) is used, the specimen is necessarily placed before the entrance slit in the full undispersed beam, and is generally imaged by the microscope objective on the entrance slit (see Seeds and Wilkins, 1950). If a line source, such as mercury, is used with a wide slit, the spectra consist of a series of monochromatic images of the specimen, and one exposure shows the absorption of various parts of a specimen (see Plate 2B opposite page 33). When a monochromator is used, it is better that the specimen should come after it, the microscope focusing the exit slit on the specimen. Such arrangements are described by Thorell (1950) and by Walker and Deeley (1955). This latter arrangement, with a double-beam photoelectric recording spectrophotometer, is illustrated in Fig. 9.5.

Quantitative measurements of absorption with a microspectrophotometer are subject to error for several reasons that are clearly discussed by Barer (1950). When the specimen is small, diffraction effects and any aberrations in the microscope will cause some radiation from the transparent surroundings of an opaque specimen to fall into the image of the latter. The magnitude of these diffraction effects depends on the focusing of the microscope, on its resolving power and aperture, and on the aperture of the condenser used to illuminate the object. The errors are eliminated where it is possible to isolate the specimen by means of an opaque disk in contact with the specimen and with a hole small enough to expose the specimen only ; the error may be estimated by replacing the specimen with an opaque disk or sphere large enough to just cover the specimen and measuring the apparent transmission. Another limitation is that, in order to obtain sufficient resolution and intensity, a microscope of large numerical aperture must be used ; the path through the specimen then becomes indefinite and absolute determination of pigment concentration inaccurate.

CHAPTER 9

Some Typical Spectrophotometers

It is proposed in this chapter to describe only a few spectrophotometers which between them will illustrate most of the principles previously discussed. It would be quite impossible to give either a complete or an up-to-date account of commercially available instruments, and those selected for description have been chosen to illustrate principles; the inclusion (or omission) of a particular instrument should not be taken as an indication of merit or lack of merit.

Atmospheric absorption

Special problems arise in regions where the atmosphere is absorbing. At $\lambda < 1800$ Å (the vacuum ultra-violet), oxygen is absorbing and vacuum instruments are used. In the infra-red, carbon dioxide and more particularly water vapour absorb in certain regions and this is dealt with in three alternative ways:

1. The instrument may be hermetically sealed and evacuated.
2. Drying agents may be used inside the instrument; or a stream of dry air or nitrogen may be passed through. If hygroscopic prism materials are used (alkali halides) the instrument must be kept reasonably dry and/or warm by means of a built-in electrical heater.
3. In the near infra-red, the absorption by water vapour is not complete in the path length in a typical instrument— see Fig. 9.1. It would be difficult if not impossible to investigate an absorption spectrum superposed on a background such as this. But where comparison is made between signals coming through two similar paths (a double-beam instrument), this water-vapour absorption will disappear from the ratio of the signals—always provided the two beams are properly 'balanced'. Most double-beam instruments can be used as single-beam instruments, to give the absorption spectrum of the atmosphere within the instrument, or the emission spectrum of the source of radiation.

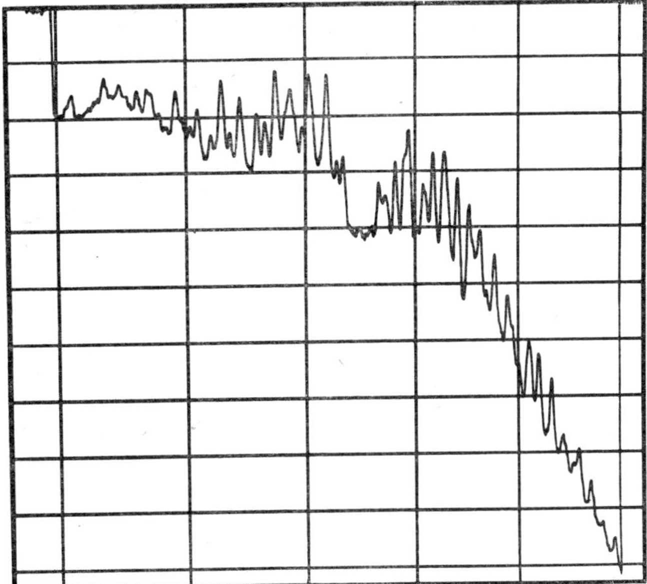

FIG. 9.1   Absorption of water vapour in 6 μ region.   Recorded on a
Beckman IR2 single-beam infra-red spectrophotometer.

SINGLE-BEAM, NON-RECORDING
ULTRA-VIOLET SPECTROPHOTOMETER

Beckman model DU, Hilger Uvispek, Unicam SP500

These three photoelectric instruments of Class 2 differ in detail
but are all similar in general design and size.   All use a 30-degree
quartz prism with an aluminized back surface and mirror focusing.
In Fig. 9.2 the monochromator slits are above and below the level
of the prism, a device that reduces aberrational errors of the
collimating mirror A.   All the instruments give readings of optical
density on the scale of a potentiometer arranged in a circuit similar
to that of Fig. 8.2.   Choice of two mounted photocells gives a
wavelength range from 2000 to 10 000 Å.   Both a hot-cathode
deuterium lamp and a tungsten-filament lamp are mounted, and
can be switched on as required.   In this class of instrument the
radiation must remain constant during each measurement and
both lamps are therefore run from circuits giving stabilized
currents.

M

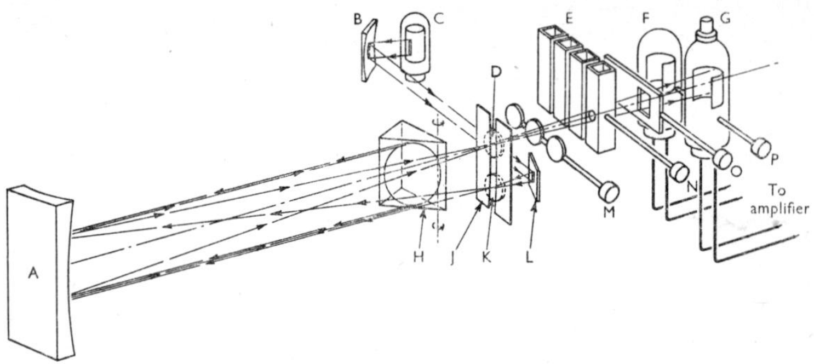

FIG. 9.2.   Schematic arrangement of Unicam SP500 photoelectric spectrophotometer.   (*A*) Collimating mirror.   (*B*) Condensing mirror. (*C*) Light source.   (*D*) Quartz lens.   (*E*) Quartz test cells.   (*F*) Ultra-violet detector.   (*G*) Infra-red detector.   (*H*) Quartz prism.   (*J*) Bilateral curved slits.   (*K*) Quartz disc.   (*L*) Plane mirror.   *(M)* Filter slide. (*N*) Test-cell slide.   (*O*) Dark-current slide.   (*P*) Detector slide.

ULTRA-VIOLET AND NEAR INFRA-RED
RECORDING SPECTROPHOTOMETERS

The Cary Model 14

This is a double-beam photoelectric recording spectrophoto-meter of Class 2 and is illustrated in Fig. 9.3.   It includes a double monochromator in which a plane reflecting echelette grating is preceded by a 30-degree quartz prism.   Radiation from a hot-cathode water-cooled deuterium lamp A or from a tungsten-filament lamp C enters the monochromator through the slit D and leaves through the slit L.   The monochromatic radiation is then sent at 30 cycles per second, via a rotating mirror O driven by a synchronous motor, alternately through the reference and specimen cells T_1 and T.   The shaft of the motor also carries a chopper disk that produces a dark interval between each half-cycle of the alternation.   The two beams are then directed to the photomultiplier, the beam from the specimen travelling via mirrors V and W and that from the reference cell via mirrors V_1 and W_1.   The photoelectric circuit is similar to that shown in Fig. 8.2.

A system of photoelectric timing signals is synchronized with the alternate pulses of specimen and reference radiation seen by the receiver, and this causes the potentiometer of Fig. 8.2 to attenuate the reference signal but not the specimen signal before they pass into the comparison circuit. When the two signals are compared, any difference causes the pen motor to further adjust the potentiometer until the signals are equal. The potentiometer is critically damped (with a period of about one second) and its movements during the equalizing of the signals are plotted against wavelength on a chart that is linear in optical density and wavelength.

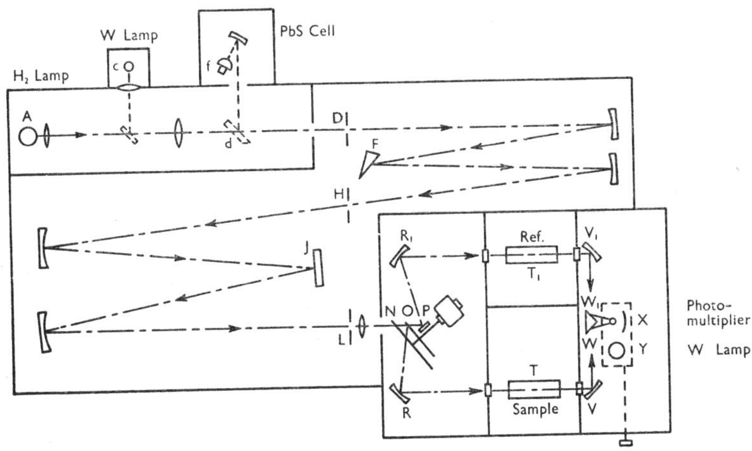

FIG. 9.3. Cary photoelectric spectrophotometer with double monochromator (prism and grating).

When the instrument is used in the infra-red region the tungsten light source Y is slid into the place of the photocell X. Radiation from this source is divided between the specimen and reference cells, chopped, and sent into the monochromator through the slit L. The pulses of monochromatic radiation emerging from slit D fall on the lead sulphide cell F. In this arrangement non-monochromatic radiation thermally emitted by the chopper unit is prevented by the monochromator from reaching the photocell. Fig. 6.6 is an illustration of a record obtained from this instrument.

Double-beam (class 1) recording infra-red spectrophotometer

In the Hilger H800 (Fig. 9.4), radiation from the Nernst lamp N

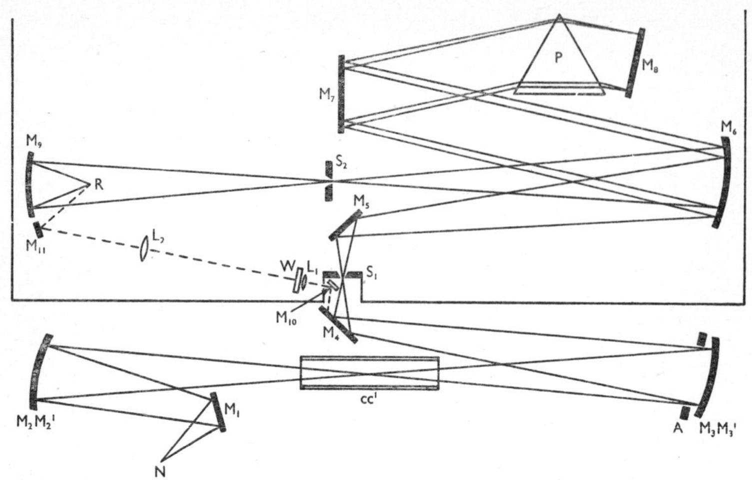

FIG. 9.4. Plan view of optical system of Hilger H800 spectrophotometer.

falls, via the plane mirror M_1, on two concave mirrors M_2 and M_2', vertically above each other. These mirrors form intermediate images of the source where specimen cells C and C' are mounted on the stand. The beams then pass to concave mirrors M_3 and M_3' which focus them, via M_4, on the slit S_1 of the spectrometer. The mirror M_4 is vibrated at $12\frac{1}{2}$ cycles per second so that the upper and lower beams are transmitted through the mono-chromator alternately. The absorbing specimen is placed in the upper beam and its absorption is balanced by the adjustable aperture A in the lower beam, the aperture opening being controlled by a servo-motor. A reduced image of the exit slit is formed on the small thermopile R. When the two beams are not equal in intensity there is an alternating current from the thermopile and this is fed through a transformer to an amplifier tuned to a narrow frequency band at $12\frac{1}{2}$ cycles per second in order to reduce the noise level; the output drives the motor until the two beams are balanced. The movement of the strip chart recorder is synchronized with the cam that rotates the Littrow mirror for changing the wave-length, and the pen is moved across the chart in sympathy with the position of the aperture A (see record in Plate 4 facing page 49). Other features of the instrument are the optional use of continuous automatic adjustment of the slit width to give approximately constant intensity, the passing of dry air through the instrument to protect the optical parts and to eliminate interference from water-vapour bands, interchangeable gratings or prisms of different materials, and a bimetallic expansion device that automatically adjusts the Littrow mirror to compensate for wavelength changes caused by temperature variations.

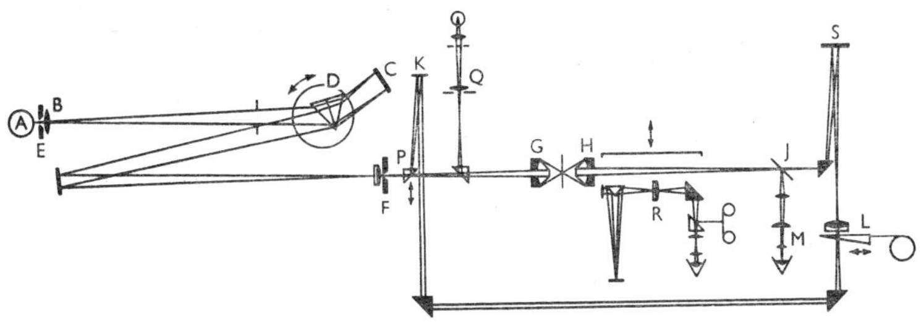

FIG. 9.5.   Double-beam photoelectric microspectrophotometer.
(From Walker and Deeley, 1955.)

Recording microspectrophotometer

Walker and Deeley (1955) describe the double-beam instrument of Class 1 that is illustrated in Fig. 9.5. Radiation from the source, a high-pressure xenon arc A, is dispersed by passing through a Littrow monochromator with 60-degree quartz prism and mirror focusing. A prism P is vibrated at 50 cycles per second, so that the monochromatic radiation from the exit slit passes to the photomultiplier alternately through the reflecting microscope GH, in which is mounted the specimen to be measured, or via the mirror K to an aperture L, which is logarithmically tapered so that its movement gives a linear change in optical density. The a.c. output of the photomultiplier S controls the position of this aperture until the two beams are balanced. The recording pen is attached to the aperture and the chart may be coupled either to the wavelength drum when a spectrophotometric curve is being traced or to the microscope stage when it is desired to measure the transmission across a specimen at a fixed wavelength. If the aperture L is to measure optical densities correctly it is essential for the illumination in its plane to be uniform; tests should be carried out and adjustments made by moving the mirror K. The size of specimen observed is about two microns. The slit width normally used corresponds to a wavelength range of 20 Å. The optical systems M and R are used for viewing and photographing respectively.

Walker and Deeley tested the performance of the microscope by using an opaque sphere as object and found that the intensity of the image, which ideally would be zero, was less than one half per cent of the full intensity.

PHOTOELECTRIC GRATING SPECTROPHOTOMETER
FOR VACUUM ULTRA-VIOLET

This is the single-beam instrument (Class 2) illustrated in Fig. 9.6. It is described by Watanabe, Inn and Zelikoff (1953) and by Inn (1955). The source of radiation is a high-voltage hydrogen discharge tube with a glass capillary Q and is placed immediately in front of the entrance slit, which has no window. The mono-chromator consists of a concave grating, of 1-metre radius, that is rotatable on the Rowland circle as described on page 160. Beyond the exit slit is an absorption cell, 47 mm long, with cemented lithium fluoride windows. The detector is a normal photomultiplier whose window is coated with sodium salicylate.

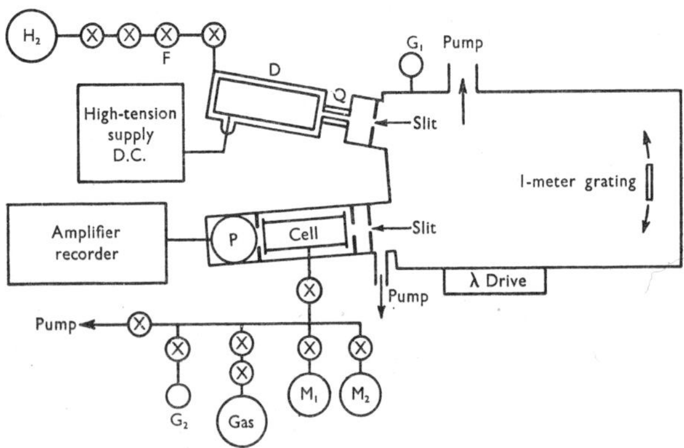

Fig. 9.6. Schematic arrangement of single-beam photoelectric spectro-photometer with concave grating in vacuum. (From Inn, 1955.)

The photoelectric output is fed to a pen recorder with a normal scanning speed of 20 Å per minute. Slit widths correspond to about 0·85 Å. The method of dealing with gas specimens is shown diagrammatically. Hydrogen diffusing through the slit from the discharge tube is removed by continuous pumping. The instrument has been used down to 1050 Å, the limit of transmission of lithium fluoride.

THE FAR INFRA-RED

There are special problems here because not much energy is available at long wavelengths. Incandescent sources approximating to a black body have a maximum emission at 1–3 μm and their emission at wavelengths longer than about 20–30 μm is very much less than this. The problem is approached in two ways: use of large diffraction gratings; or, more recently, an entirely different principle has been developed—Fourier transform spectroscopy. These will be described briefly in turn.

DIFFRACTION GRATING INSTRUMENTS

In the use of diffraction gratings, it is not difficult to obtain the large instrument needed because the coarse gratings suitable for wavelengths longer than about 100 μ can be made in large sizes. McCubbin and Sinton (1952) have made the instrument illustrated in Fig. 9.7. Radiation entering the slit is reflected from the parabolic mirror (diameter 1 ft and aperture F/1) to the plane echelette grating (diameter also 1 ft). Diffracted radiation is reflected back to the Golay pneumatic detector, whose entrance window serves as the exit slit. The quartz lens that focuses the radiation on the slit also acts as a filter to remove wavelengths to which it is opaque (4–40 μm) and wavelengths below 4 μm are largely removed by focal isolation—the chromatic aberration of the lens is so great that the short waves are not focused on the slit. (Quartz has refractive indices of 1·4 at 2 μm and 2·2 at 100 μm.) The gratings used are ruled in a milling machine with either 67 or 25 lines per inch, and the resolution obtained is about 1 cm^{-1}. The most satisfactory source of radiation for this region is a high-pressure mercury arc in quartz; the long-wave radiation is partly from the mercury vapour and partly from the hot quartz.

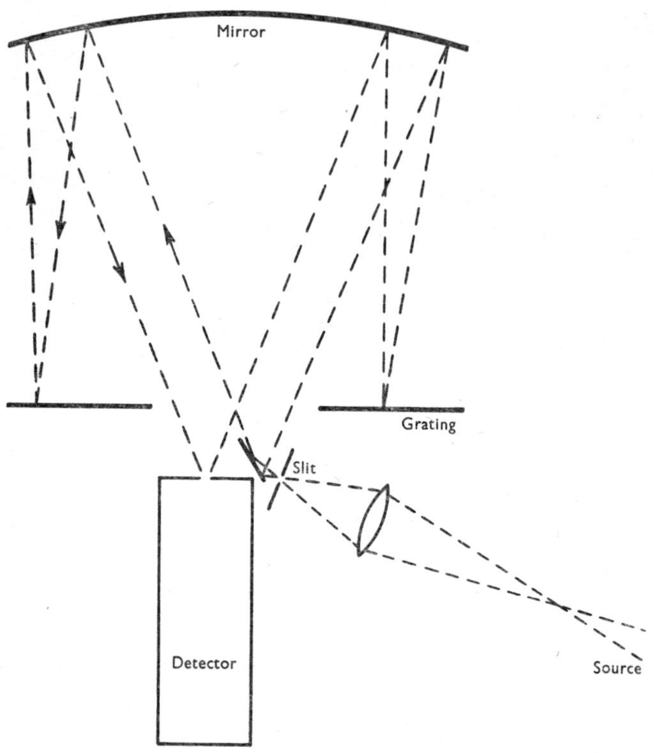

FIG. 9.7.   Infra-red spectrophotometer with coarse echelette grating. (After McCubbin and Sinton, 1952.)

FOURIER TRANSFORM SPECTROSCOPY

The method now to be described has been largely developed since about 1957. It is alternatively named Fourier transform spectroscopy, interference spectroscopy or Michelson spectroscopy. The subject is reviewed in some detail by Jacquinot (1960) and by Gebbie and Twiss (1966).

The method is based on the interference fringes produced

with a Michelson interferometer. This consists essentially (Fig. 9.8) of two fully reflecting mirrors M_3 and M_4 and a diagonal plane B which partly transmits and partly reflects. There is a common beam from the source to the diagonal plane B from which radiation can reach the Golay detector by alternative paths $BM_7BM_8M_3$—and $BM_4BM_8M_7$—of lengths x_1, x_2, respectively.

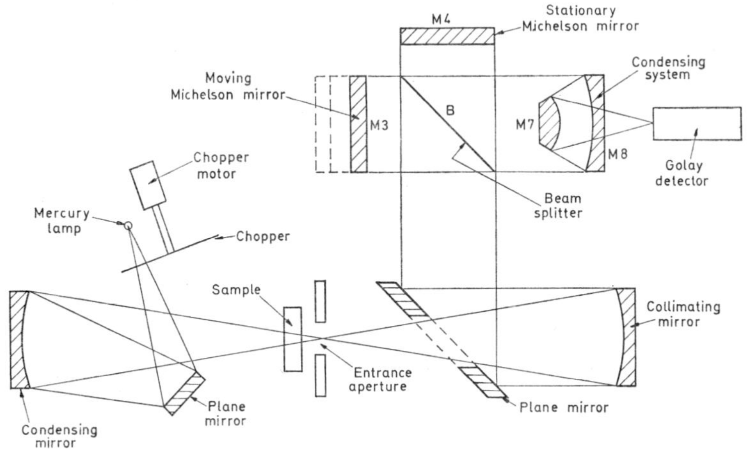

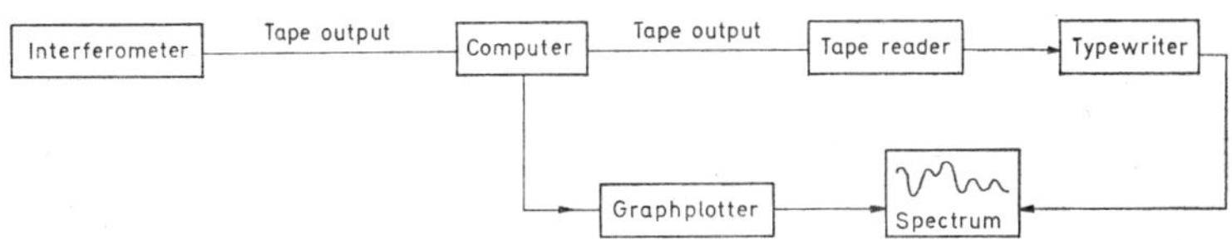

FIG. 9.8.   Interferometric spectrometer (after Grubb-Parsons).

The two wave trains reaching the detector will interfere; if their path difference is a whole number of wavelengths, $n\lambda$, a maximum amplitude will be obtained; but if the path difference is $(n+\tfrac{1}{2})\lambda$, the resultant amplitude will be a minimum.   The equation

$$y = a \sin 2\pi(\nu t - x/\lambda)$$

represents, at time t and position x, the displacement y in a wave of amplitude a, frequency ν and wavelength λ, travelling along the x direction. (See textbooks on physical optics or general wave-motion.) The interference of two wave trains can be represented by the addition of two such equations with different x values, x_1 and x_2; the addition can be transformed into a product by a standard trigonometric relation. Thus

$$y = y_1 + y_2 = a \sin 2\pi(vt - x_1/\lambda) + a \sin 2\pi(vt - x_2/\lambda)$$
$$= 2a \cos [\pi(x_1 - x_2)/\lambda] \cdot \sin 2\pi[vt - (x_1 + x_2)/2\lambda]$$
$$\longleftarrow\!\!\text{---amplitude---}\!\!\longrightarrow \; . \; \leftarrow\!\text{periodic wave motion}\!\rightarrow$$

The right-hand sine term represents a wave motion and the left-hand term is its amplitude. The intensity I is the square of the amplitude. Thus writing the path difference $x_1 - x_2$ equal to δ we have,

$$I = 4a^2 \cos^2 (\pi\delta/\lambda)$$
$$= 2a^2 + 2a^2 \cos (2\pi\delta/\lambda) \tag{9.1}$$

This resultant intensity varies periodically with δ, as in Fig. 9.9a. If the radiation is perfectly monochromatic (an unattainable ideal) the intensity will vary periodically in this manner for path differences up to infinity. But if (to take another ideal case) it consists of two perfectly monochromatic radiations of slightly different wavelengths λ_1 and λ_2, the path difference may at the same time be $n\lambda_1$ and $(n + \frac{1}{2})\lambda_2$; thus a maximum intensity for one wavelength will be superposed on a minimum for the other wavelength. If the two wavelengths are present in equal intensities the fringes will disappear (Fig. 9.9b) to reappear for a greater path

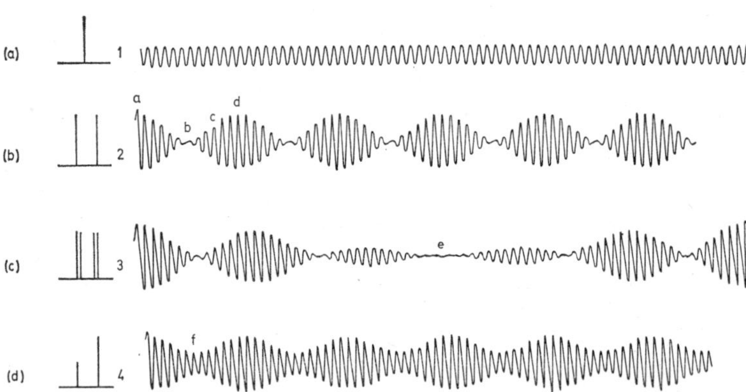

FIG. 9.9.   Interferograms for the four simple spectral distributions indicated at the left. (Reproduced in various textbooks—original in Michelson, *Light waves and their Uses*, 1903, University of Chicago Press.)

difference. The variations of the interference pattern with path difference for other simple spectral distributions are shown in Fig. 9.9c and d. This figure is reproduced from Michelson's work, who in 1891 used the method to deduce the structure of spectral lines which could not at that date be resolved. With the advent of computers the principle can be extended to spectral distributions which are not simple. To develop the theory:

Suppose we have a source of radiation whose intensity at wavelength λ is I_λ, then the resultant intensity for a path difference δ is the sum of the terms of expression (9.1) for all wavelengths, i.e.

$$I_\delta = \int_{\lambda=0}^{\lambda=\infty} I_\lambda \, . \, d\lambda + \int_{\lambda=0}^{\lambda=\infty} I_\lambda \cos\left(2\pi\delta/\lambda\right) . \, d\lambda \qquad (9.2)$$

The first term is a constant independent of the path difference and the second term varies with δ in a manner depending on the variation of I_λ with λ. Fig. 9.10a is a typical curve showing the relation between I_δ and δ for a source having a relatively complicated variation of I_λ with λ.

It can be shown (Fourier integral theorem) that I_λ can be expressed as the Fourier transform of I_δ, i.e.

$$I_\lambda = \int_{\delta=0}^{\delta=\infty} I_\delta \cos\left(2\pi\delta/\lambda\right) . \, d\delta \qquad (9.3)$$

With the aid of a computer this integration can be carried out numerically by using a sufficient number of ordinates from an interference curve.

In Fig. 9.8, the Michelson interferometer proper has already been described. The radiation source, which is chopped to permit a.c. amplification, is imaged by a condensing mirror on an aperture where an absorbing sample may be placed. The beam is then rendered parallel by a collimating mirror, and after passage through the interferometer itself it is focused on the Golay detector by mirrors M_8, M_7. The diagonal plane B is a plastic film (often I.C.I. Melinex: polyethylene terephthalate) and the source is a high-pressure mercury lamp. To avoid interference due to atmospheric absorption the whole optical equipment should be in vacuum, or kept with dry nitrogen.

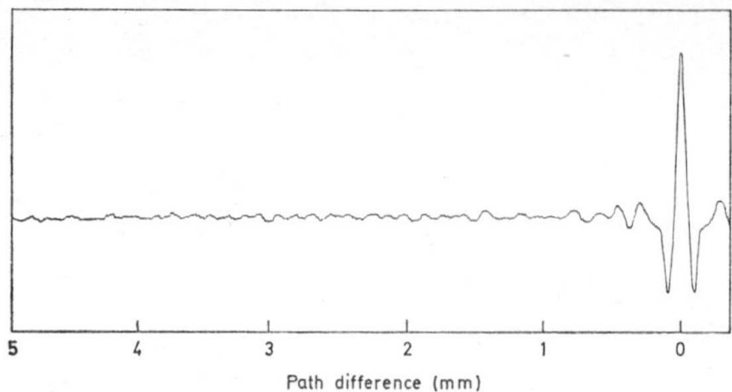

FIG. 9.10*a*. Part of an interferogram produced with a Michelson interferometer using an SiO_2/Hg source, mirrors of 20 cm diameter and a Golay detector. The speed of the movable mirror was 150μm/min and the range of path difference (twice the mirror movement) was 0-5 cm. The figure shows only path differences 0-5 mm. (Reproduced from the originals obtained at The Physics Department, University of Exeter).

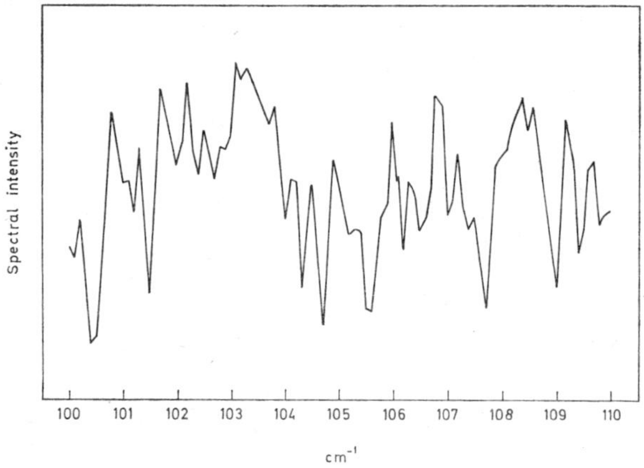

FIG. 9.10*b*. Part of the Fourier transform of the interferogram, covering the range $\bar{\nu} = 100\text{--}110$ cm⁻¹ showing the emission spectrum of the source as modified by absorption of the atmosphere within the apparatus. The ordinates of Fig. 9.10*a* were sampled every 20·3 μm and fed to an Elliott 803 computer. The resolution of the spectrum should be 1/(maximum path difference); i.e. 0·2 cm⁻¹. Points for Fig. 9.10*b* were computed and plotted at half this interval.

Fig. 9.10 shows an interferogram obtained and the spectral distribution of the source calculated from this. It is not strictly necessary to plot the interferogram. As the mirror M_3 is traversed the amplified output voltage from the detector can be punched direct on computer tape. The computer answer can be provided as a set of figures or as a plotted curve showing either I_λ versus λ, or as the ratio of two I_λ curves, i.e. a transmission curve.

The advantages of this seemingly complicated method are:

1. The source, effectively the entrance aperture, can be circular, which permits a higher intensity of radiation to be used than is possible when the source has to be a narrow slit as with a diffraction grating.

2. When, as in the far infra-red, the energy available is limited, the performance is limited by noise in the detector (see page 162). In interference spectroscopy, the whole spectrum falls on the detector at one time, so that the signal/noise ratio is enormously increased.

Equation (9.3) calls for integration from $\delta = 0$ to $\delta = \infty$. In practice, some maximum path difference δ must be imposed by the limited traverse of the mirror M_3, and this limits the resolution. But some improvement can be effected by modifying the computing with an additional term, often named a weighting or apodization function.

To compare the limiting performances of Fourier transform and dispersion (grating) spectroscopy: in the Fourier method, to double the resolution we must double the path difference which doubles the time of measurement. But to double the resolution with a grating (or prism) instrument, we must halve the width of each slit, thus reducing the signal to one quarter. When the detector is working at the noise limit this means increasing the observation time by a factor of 16 [compare equation (4.8) on page 73]. This aspect is discussed by Richards (1964).

REMOVAL OF STRAY RADIATION

As already emphasized, this is extremely important for the far infra-red and it will be useful to summarize the various methods, some or all of which may be used in any one instrument:

1. *Focal isolation* as with the quartz lens mentioned on page 183.
2. *Use of absorbing filters.* A quartz lens or quartz plate

absorbs the region 4–40 μm. Other materials absorbing short wavelengths are shown in Table 9.1.

TABLE 9.1

Material	Short-wave absorption edge (in microns)	Notes
Germanium	2·0	Used as a thin film by Brattain and Briggs (1949)
Bismuth	2·0	Plyler and Bell (1952)
Antimony	2·0	Plyler and Bell (1952)
Selenium	2·0	Transmits to 18 μ, but with some absorption at 12 and 14 μ (Gebbie and Cannon, 1952)
Lead sulphide	2·9 ⎫	Used as a thin evaporated film, coated with selenium to reduce interference by internal reflections (Braithwaite, 1955)
Lead telluride	3·8 ⎬	
Lead selenide	4·2 ⎭	

Crystals of alkali halides that are normally transparent in the infrared can develop absorption bands when they contain certain impurities, such as additional alkali metal; this is caused by the presence of so-called F-centres (from the German *Farbzentren*). How far these bands extend into the infra-red depends on the concentration of the impurity. Typical transmission curves for such filters are shown in Fig. 9.11.

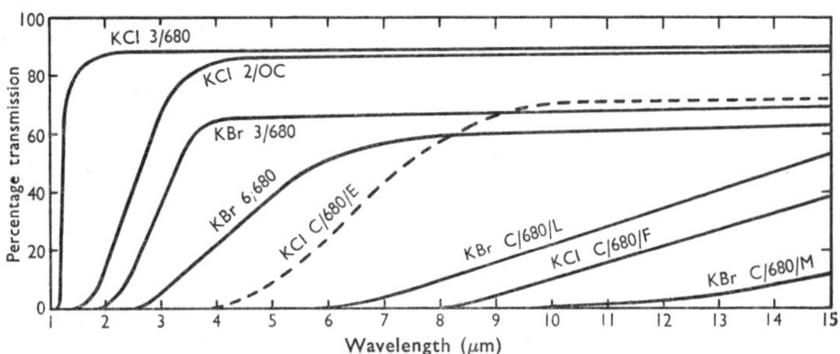

FIG. 9.11. Infra-red transmissions of some F-centre filters.

3. *A rotating chopper* made from one of the materials shown in Figs. 4.4 and 4.5 (page 66) will modulate the long waves to which they are opaque, but short waves to which they are transparent will be unmodulated; and the d.c. signal at the detector due to the short wavelengths will not be amplified.

4. *Reststrahlen* ('residual rays'). Materials which absorb very strongly also exhibit high reflection, e.g. metals which are still opaque in very thin films. Thus a plate of quartz, which shows intense absorption in the range 5–40 μm can be used to reflect these wavelengths along the required path, while the longer and shorter wavelengths are freely transmitted. Other materials used to remove short wavelenths in this way are (Fig. 4.5):

when working in the range 40–70 μm,　　　　　　　NaCl

when working in the range 110–150 μm,　　　　　KRS5

5. *Scatter.* Finely divided matter scatters short wavelengths much more than long wavelengths (Rayleigh's $1/\lambda^4$ law. Thus a thin layer of soot or MgO deposited on a transparent plate or film is often used; see Plyler and Bell (1952).

6. *A diffraction grating as a filter.* When the spacing (a) of a grating is less than half a wavelength, no diffraction into any non-zero order can occur. In the grating equation

$$n\lambda = a(\sin \phi \mp \sin \theta)$$

this would mean that the term in brackets would be greater than 2. Thus all the radiation must go into the zero order, i.e. be specularly reflected. But shorter wavelengths can be diffracted out of the beam into non-zero orders. This method is discussed by White (1947).

SPECTROFLUORIMETERS

Alternative names for spectrofluorimeters are spectrofluorometers or spectrofluorophotometers. A spectrofluorimeter should include (see Fig. 9.12):

1. A radiation source S, which is usually either a high-pressure mercury lamp with a number of intense lines in the ultra-violet or a xenon lamp which has an intense continuous spectrum.

2. A monochromator M_1 to isolate a spectrum line or band.

3. A specimen holder H which, in the case of liquids, may be a cell with clear faces perpendicular to each other. A number of such cells can be mounted on a slide or rotatable turret. Alternatively, opaque specimens may be mounted as shown in H' in the figure.

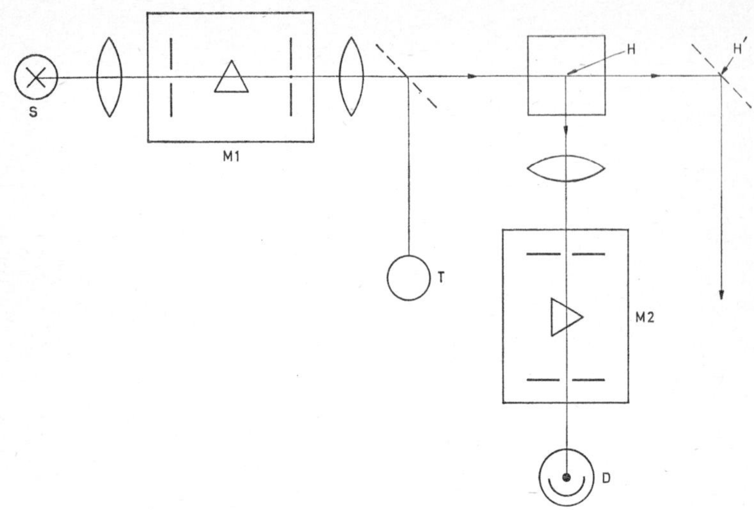

FIG. 9.12. Block diagram to show the principle of a spectrofluorimeter: *S*—source of radiation; *M*1, *M*2—monochromators; *H*—Fluorescent specimen; *H'*—Reflecting fluorescent specimen; *D*—Detector (photomultiplier) of fluorescence; *T*—Detector of exciting radiation.

4. The fluorescent radiation may be condensed into a second monochromator M_2, whose output may be fed to . . .

5. A radiation detector D. A photomultiplier is most commonly used, the output being fed to an amplifier and possibly to a pen recorder. This last can then be made to plot fluorescent intensity against the wavelength setting of M_1 (the excitation spectrum) or against the setting of M_2 (the fluorescence spectrum). Such curves are illustrated in Fig. 6.2 on page 110.

Either of the monochromators may be omitted and replaced by a filter. If both monochromators are omitted the instrument becomes a simple 'fluorimeter'; a number of such instruments have been available commercially for many years. At the time of writing, several complete spectrofluorimeters are also available.

Excitation spectrum

If the fluorescence intensity is observed at a single wavelength, and the exciting wavelength (setting of M_1) is varied, the resulting 'excitation spectrum' curves (left-hand curves of Fig. 6.2) have ordinates proportional to the energy absorbed by the fluorescent specimen; i.e. the curves are analogous to the distorted absorption curves obtained with a single-beam infra-red spectrophotometer. The distortion arises because of the wavelength variation of

(a) the intensity of the source and
(b) the transmission and dispersion of the monochromator M_1.

This distortion may be corrected by monitoring, at T in Fig. 9.12, a portion of the radiation from M_1. T should be a detector for which either the energy response or the quantum response is independent of wavelength. Devices used at T are

(a) a thermopile or
(b) a fluorescent specimen which is completely opaque to all wavelengths from M_1. Behind T is then placed a photocell, and the response of this will be proportional to the radiation incident on T. A suitable specimen that has been used by Parker and Rees (1958) is a solution of fluorescein ($4 \cdot 4 \, . \, 10^{-3}$ M in $0 \cdot 1$-N $Na_2CO_3/NaHCO_3$).

If the complete spectrofluorimeter is then made to record log (response of T)/log (response of D), the result is a corrected excitation curve whose ordinates are proportional to absorbances of the fluorescent solution; to quote Parker and Rees (1962), 'the measurement of true excitation spectra thus provides a method of measuring absorption spectra of fluorescent compounds at concentrations far lower than would be needed to measure the absorption spectra directly.'

Fluorescence spectrum

If the wavelength setting of M_1 (Fig. 9.12) is kept fixed and M_2 is made to traverse the spectrum, the curve traced will be the 'fluorescence spectrum' of the fluorescing specimen (right-hand curves of Figs. 6.2, page 110). These curves will be distorted with a simple instrument, this time because of wavelength variations of

(a) transmission and dispersion of M_2;
(b) sensitivity of the detector D.

N

The necessary corrections may be determined in preliminary tests with a spectrofluorimeter by recording the ratio of response at T with a ' neutral detector ' as above to the response at D, without a fluorescing specimen at H. The two monochromators are made to traverse the spectrum together. The results of this calibration may be incorporated into the instrument—either as a programmed slit variation or as a programmed potentiometer in the output.

Standard fluorescence specimen

A useful alternative to calibrating an instrument in the way just described is to compare unknown specimens with a standard fluorescent material. Quinine sulphate dissolved in 1-N H_2SO_4 is commonly used for this purpose. Melhuish (1960) has investigated this material carefully and reports that the shape of

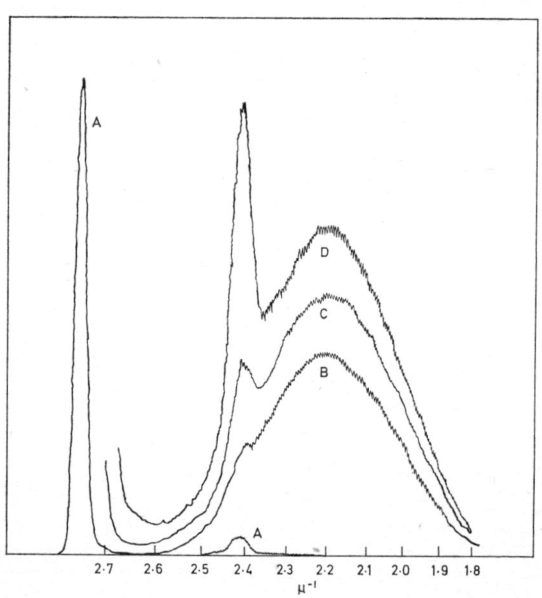

FIG. 9.13. Fluorescence spectrum of quinine sulphate in 0·1-N H_2SO_4 excited by $\lambda = 365$ nm. Concentrations of quinine sulphate: A, zero, B, 0·1 μg/ml; C, 0·033 μg/ml; D, 0·01 μg/ml. The peak at 2·75 μm^{-1} is Rayleigh scattering of the exciting radiation; and the peak at 2·4 μm^{-1} is Raman scattering of the exciting radiation by water molecules
(Reproduced from Parker, C.A., 1959)

its emission curve is independent of concentration in the range $5 \cdot 10^{-3}$–$5 \cdot 10^{-5}$ M; the quantum efficiency is high, namely 0·52; this means that for every 1000 photons absorbed by the solution, 520 fluorescence photons are emitted somewhere within the fluorescence spectrum.

Fig. 9.13 shows the fluorescence spectrum of quinine sulphate solution covering the range $\lambda = 4000$–5500 Å ($\bar{v} = 2 \cdot 5$–$1 \cdot 8\ \mu m^{-1}$). At the smallest concentration (curve D), $c = 0 \cdot 01\ \mu g/ml$, the 'noise' of the spectrofluorimeter shows up in the thickness of the line. This thickness indicates a limiting detectable concentration of about $0 \cdot 0003\ \mu g/ml$. To detect this concentration by its absorption spectrum would require, assuming $\epsilon_{mol} = 10^4$ and $\Delta A = 0 \cdot 002$, a cell length of no less than 300 cm.

Superposed on the fluorescence spectrum is a Raman line of water at $\lambda = 4100$ Å ($\bar{v} = 2 \cdot 4\ \mu m^{-1}$). This line would not interfere seriously in the estimation of quinine sulphate, provided a wavelength away from this region was chosen for measurement.

ABRIDGED SPECTROPHOTOMETERS

These instruments eliminate the need for a monochromator by using either a continuous spectrum or line spectrum (mercury lamp) source with filters transmitting chosen wavebands. Or in the Luft instruments described below, the detector responds only to a certain waveband.

One of the many filter instruments available commercially is the Spekker absorptiometer.

The Spekker absorptiometer

This is a single-beam instrument using the principle of optical balancing—Class $1a$. Its optical system is illustrated in Fig. 9.14. A lens K, mounted to the right of the lamp G forms a parallel beam of light that passes on through the specimen in the cell L to another lens N. This lens forms an image of the lamp filament on a photo-voltaic cell O (the indicating cell) mounted on the right of the instrument. The amount of light reaching this photocell is controlled by a variable shutter J. The shutter aperture is controlled by a calibrated drum that allows the intensity of the light to be varied by known amounts.

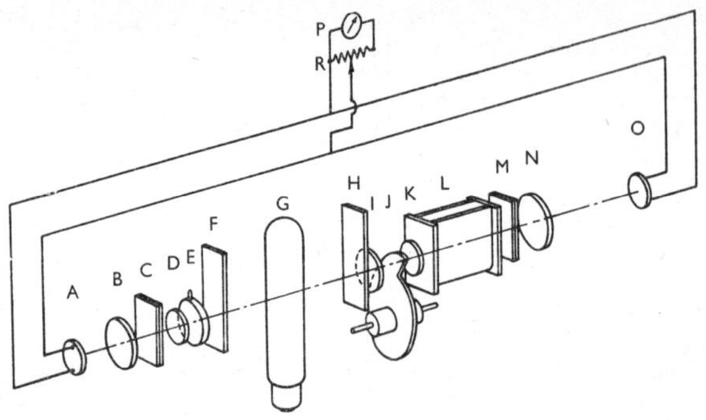

FIG. 9.14. Schematic arrangement of Spekker absorptiometer. (*A*) and (*O*) Photocells. (*B*) Window. (*C*) Spectrum Filter. (*D*) lens. (*E*) Iris diaphragm. (*F*) Heat-absorbing filter. (*G*) Lamp. (*H*) Heat-absorbing filter. (*I*) Window. (*J*) Variable shutter. (*K*) Lens. (*L*) Cell for liquids. (*M*) Spectrum filter. (*N*) Lens. (*P*) Galvanometer. (*R*) Sensitivity control.

Light from the lamp also falls on a similar photocell *A*, mounted on the left of the lamp. This is known as the compensating cell, and the amount of light falling on it can be varied by means of an iris diaphragm *E*. The two photocells are connected in series, and in circuit with a galvanometer, in such a way that when the photoelectric currents given by the cells are equal the galvanometer shows no deflection. Colour filters may be inserted in each beam to give wavebands 200–400 Å wide. When a mercury lamp is used with suitable filters in the instrument, truly monochromatic radiation may be obtained, including the ultra-violet line at 3650 Å.

The instrument is set to a null reading with the absorbing specimen and ' absorbing ' shutter *J* in turn.

Selection by detectors—the Luft method

Another method of isolating a required waveband is to use detectors that respond only within narrow limits. Such a method has been devised by Pfund (1940) and Luft (1943) and can best be explained by reference to its practical application in the infrared gas analyser (see Fig. 9.15). Infra-red radiation from hot filaments at *A* passes through two adjoining cells at *B*—one filled with air and the other with air containing a small quantity of the gas to be detected (CO_2 for example). The two beams then pass into two chambers at *C*, filled with the gas to be detected. These two

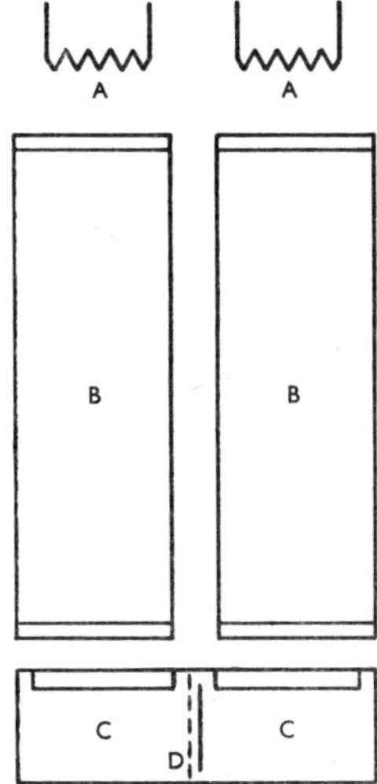

FIG. 9.15.   The infra-red gas analyser.

chambers are separated by a thin metal diaphragm D, which forms one plate of a parallel plate condenser. The gas in the two receiver chambers absorbs energy at certain wavelengths (principally 4·3 μm in the case of CO_2), thus increasing the pressure on each side of the diaphragm. If some of the absorbing gas is also present in one of the cells at B, some of the radiation is absorbed by it and the corresponding chamber at C suffers less pressure change. This moves the condenser diaphragm D and a suitable electric current can be used to measure the resulting change of capacity. By using cells 30 cm long it is possible to detect about 0·001 per cent of carbon dioxide in air. The beauty of the arrangement is that impurities other than CO_2 that absorb at different wavelengths will have no effect on the 4·3 μm radiation absorbed by CO_2. The instrument is also sensitive enough to detect about 0·001 per cent nitrous oxide (N_2O), which is not normally present in air and is harmless in low concentrations ; this detection is therefore of use in testing ventilation systems, vacuum chambers, etc. The instrument can also be used in the detection of methane (CH_4) in mines and gives a sensitivity of about 0·005 per cent.

Luft (1954) has described an improvement on the original design, and has applied the principle of selection in a somewhat different way. An internal filter, consisting of a wedge of the material to be tested, is oscillated in the beam ; only the absorbed wavelengths are modulated and the receiver therefore detects only these. This arrangement can give high sensitivity and has been used to detect 10^{-3} per cent of cyclohexane in benzene.

Wright and Herscher (1946) have used the principle of selection to estimate ethyl benzene by means of its ethyl-group absorption band at $3 \cdot 4 \ \mu$m. Their arrangement consists of a filament source, a specimen containing ethyl benzene for analysis, and a receiver consisting of ethyl-cellulose in contact with a bolometer wire.

PHOTOGRAPHIC SPECTROPHOTOMETRY

In these days, the photographic method is used only for special purposes, e.g. when it is required to photograph in a short time the spectrum of a material that is unstable. For this reason, there is little point in describing any particular instrument. To expand a little the 'principles' of photographic photometry discussed in Chapter 4 (page 79):

To obtain a value for absorbance, the blackening produced on a plate by an absorbing specimen at some wavelength must be compared—preferably with the aid of a microphotometer—with the blackenings produced on the same plate by a series of standard exposures. These may be obtained by:

1. Exposures through a calibrated adjustable aperture, similar to that used with the infra-red instrument shown in Fig. 9.4. A form of photographic instrument using such an aperture is the Spekker ultra-violet spectrophotometer; the series of exposures with this instrument is illustrated in Plate 3 and the resulting absorption spectrum obtained from the match points on these exposures is shown in Fig. 9.16. This instrument requires a number of exposures through the absorbing specimen, so that in fact it is of no use with unstable materials.

2. A stepped filter placed close in front of the spectrograph slit will give different exposures at different slit heights. A suitable filter consists of a series of platinum coatings on a quartz

plate, each step being of height 1 mm; the absorbances of the steps might be 0, 0·2, 0·4, 0·6, ... 1·8. A modification of this is a wedged filter in which the absorbance varies continuously with height. Such filters are reasonably ' neutral ' but the absorbance does vary a little with wavelength and for accurate work a calibration at a number of wavelengths may be desirable.

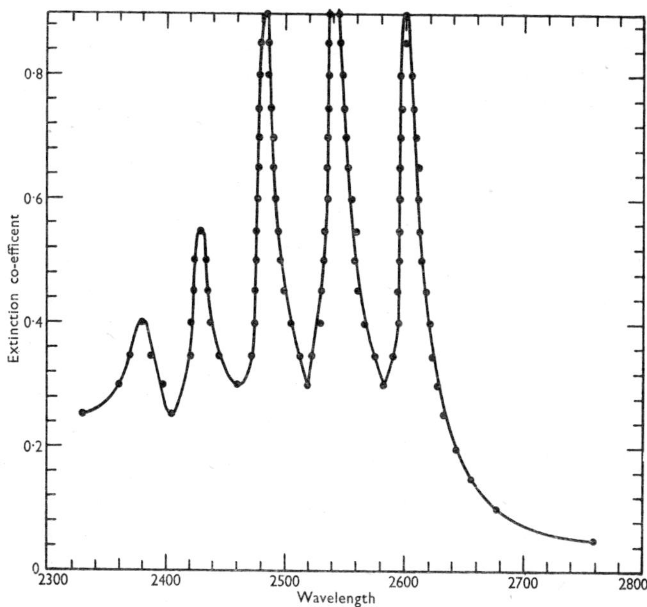

FIG. 9.16.   Absorption curve for benzene constructed from spectrogram shown in Plate 3.

3. An alternative to the stepped or wedged filter is a rotating sector wheel whose absorbance, determined by the ratio of open and opaque angles, varies with radial distance from the centre of rotation. Care must be exercised in using this, since the exposure of the test specimen is continuous and that of the sector wheel is intermittent; the conditions under which there is no ' intermittency effect ' need to be carefully tested.

VISUAL SPECTROPHOTOMETERS

These are likewise hardly used today and no particular instrument merits special description. In all visual instruments there is a field of view consisting of two juxtaposed halves whose brightness is to be matched by eye. The brightness of the adjustable half may be controlled by an adjustable aperture, a pair of polarizing prisms or polaroids, or a rotating sector.

MEASUREMENT OF DICHROIC ABSORPTION WITH PLANE POLARIZED RADIATION

In the ultra-violet and visual region, plane polarized radiation may be produced by using one of the well-known devices: a Nicol, Foucault or Rochon prism. The last named produces two polarized beams plane polarized at right angles, and one must be trapped by means of a suitable stop; or, using the phenomenon of dichroism itself, a ' polaroid ' filter may be used.

In the infra-red, it is usual to produce plane polarization by transmission through a ' pile of plates ', the radiation being incident at or near the polarizing angle. Materials must be transparent for the required wavelength range and, if highly efficient polarization with few plates is required, should have a high refractive index. Very thin films of selenium may be obtained by vacuum evaporation on to a support that is afterwards dissolved away (Elliott, Ambrose and Temple, 1948). The selenium may also be evaporated on to collodion film for use up to 40 μm (Barchewitz and Henry 1954). Conn and Eaton (1954b) have used selenium on rocksalt plates and have found that, if the plates are mounted not less than 0·25 in. apart to eliminate the disturbing effects of multiple reflection, a polarization of 99·8 per cent may be obtained with eight films.

Newman and Halford (1948) have used sheets of silver chloride in the same way. A pile of six films gives more than 90 per cent polarization in the 2–20 μm range. As a polarizer, silver chloride has the advantage that it may be prepared from sheets, 0·002 in. thick, that are available commercially ; they must not be exposed to daylight.

THE MEASUREMENT OF CIRCULAR DICHROISM

This can be done with a unit between the exit slit of the mono-chromator and the photocell in an instrument of the type shown in Fig. 9.2. The unit consists of:

1. A polarizing prism R (Fig. 9.17), which may be a Rochon prism, one of the two plane polarized beams produced being lost at a stop as shown.

2. An arrangement to convert this plane polarized radiation at will to d- or l-circularly polarized radiation. This involves producing a relative phase change of 90° between two equal plane polarized beams, polarized in mutually perpendicular planes, which are equivalent to the plane polarized beam from R. (This is explained in textbooks of optics.) Such an arrangement is in effect a quarter-wave plate. It may take several alternative forms:

   (*a*) A Fresnel rhomb (*F*, Fig. 9.17) which produces the phase change by means of two total reflections. The emergent circular polarization may be made d- or l- by rotating the Rochon prism R through 90°.

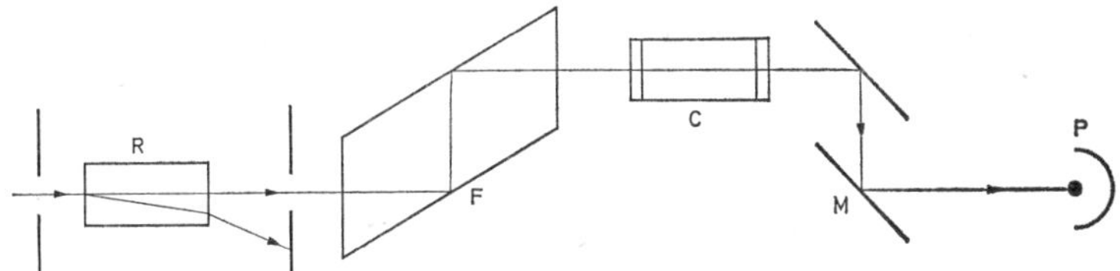

FIG. 9.17. Attachment for measurement of circular dichroism. R. Rochon prism. *F*, Fresnel rhomb. *C*, absorption cell. *M*, pair of mirrors. *P*, photocell.

   (*b*) A crystal of ammonium dihydrogen phosphate (ADP). This becomes doubly refracting when an electric field is applied across it. The phase change produced in an incident plane-polarized beam may be made 90° by adjustment of the voltage; and the sense of the circular polarization is reversed (i.e. made d- or l-) by reversing the voltage. An instrument using this arrangement is the Roussel-Jouan Dichrograph in which an alternating voltage (50 c/s) is applied. Thus the instrument measures the d- and l- absorbances every 1/50 of a second; it is in fact arranged to record directly the disymmetry factor g' of equation (6.6), page 127.

(c) A relatively thick doubly refracting plate which gives a 90° phase retardation at a number of closely spaced wavelengths; and so, on traversing the spectrum, l- and d-circular polarization is produced every, say, 80 Å.

3. The specimen cell C.

4. A pair of mirrors M to return the beam to its original axis; these are required only when the Fresnal rhomb is used.

More complete accounts of such apparatus are given by Velluz, Legrand and Grosjean (1965) and Woldbye in Chapter 5 of a volume on the uses of circular dichroism edited by Snatzke (1967).

An alternative method of obtaining a value for $(\epsilon_l - \epsilon_d)$ depends on the fact that 'after plane polarized light has passed through an optically active medium in a region of electronic absorption, the circular components have been unequally absorbed so that on recombination of the unequal components, elliptically polarized light is obtained instead of plane polarized light' [Gillard (1963)]. $\epsilon_l - \epsilon_d$ may then be deduced by measuring the parameters of the elliptically polarized radiation using a standard method [see e.g., Longhurst (1967), p. 502].

CHAPTER 10

Testing Spectrophotometers
Window Materials and Solvents

ADEQUATE testing of a spectrophotometer gives the user confidence in results obtained from it. Tests should include checking both wavelength and absorbance scales.

To ensure that an instrument is working satisfactorily it should be tested under conditions as stringent as those under which it will be operated. For example:

1. If the instrument is to be used at a wavelength near the limit of the sensitivity of the photocell, a test should be made near that wavelength. There is more chance of errors in measurements here, because the photoelectric current will be smaller and stray radiation may be greater. If, for example, measurements are required in the region 2000–2500 Å, tests might be made using astraphloxin perchlorate (see page 209).

2. Tests should be made on a specimen at a sharp maximum or minimum of absorption in order to test fully the performance of the spectrometer. Thus, measuring on a specimen with a flat maximum (such as alkaline potassium chromate at $\lambda = 3715$ Å) would give correct results even if the spectrometer were slightly out of adjustment for focus or wavelength, but tests on a sharp peak would give a low absorbance reading. (One would, of course, have to take account of the effect of finite slit width as discussed on pages 24 et seq.)

STANDARD WAVELENGTHS

These include both emission lines and absorption bands. In 1961 the International Union of Pure and Applied Chemistry recommended a set of lines and bands suitable for calibrating, in the region 2·5–17 micro-metres, i.e. $\bar{v} = 4000 - 600$ cm^{-1}. The recommendation is published by Butterworths under the title *Tables of*

Wavenumbers for the Calibration of Infra-red Spectrometers. For all the spectral regions, suitable calibration wavelengths can be chosen from data given in the various published atlases of absorption bands, some of which are listed under Further Reading on page 215.

Deuterium lines

The deuterium lamps included in so many instruments for the visible and ultra-violet regions emit the Balmer lines of the atomic spectrum in addition to the continuum for which they are primarily used. Wavelengths for these lines, and the slightly different values for a hydrogen lamp are:

Deuterium	6561·0 Å	4860·0 Å
Hydrogen	6562·8 Å	4861·3 Å

Mercury lines

The mercury spectrum includes a number of lines in the ultra-violet, visible and near infra-red. The wavelengths are listed below because many books of tables give only a few of the lines; a useful table is given, however in *International Critical Tables*, Vol. 5, page 299 (1929, first edition, McGraw Hill). In the following list, lines which are in many lamps particularly strong are given in bold type, though the distinction between strong and not so strong lines is naturally somewhat arbitrary.

Mercury wavelengths (Å)

15 295	**4358**	3021
13 950	4347	2967
13 673	4339	2894
13 570	4078	2848
11 888	**4047**	2803
11 287	3984	2753
10 140	3906	2699
6907	**3663**	2652
5791	**3655**	**2537**
5770	**3650**	2535
5461	3341	2483
4960	**3132**	2399
4916	3126	

Other useful emission lines are

Krypton (Acquista and Plyler, 1952): 1·278, 1·321, 1·374, 1·392, 1·442, 1·475, 1·533, 1·686, 1·817 μm.
Potassium: 0·7765, 1·1773 μm.
Rubidium: 0·7800, 0·7948, 1·4754, 1·5929 μm.
Caesium: 0·8521, 0·8944 μm.

Absorption maxima recommended by Acquista and Plyler (1952) for calibration up to 2·6 μm are

Didymium glass: 0·684, 0·740, 0·748, 0·808, 0·880, 1·067, 1·220, 1·517, 1·918 μm.
Polystyrene film: 1·681, 2·144, 2·170, 2·187 μm.

They also recommend the use of some absorption maxima in the spectrum of 1 : 2 : 4-trichlorobenzene, but the spectrum is relatively complex and the chosen maxima can best be identified by consulting the spectrogram in their paper. Mecke and Oswald (1951) give lines, covering the range 0·85–2·4 μm, that they have found useful in the absorption spectra of chloroform, trichlorethylene and benzene.

Downie, Magoon, Purcell and Crawford (1953) list absorption bands, together with the emission lines of mercury, suitable for the calibration of instruments with prisms of SiO_2, LiF, CaF_2, NaCl, KBr, and CsBr. The selected bands are from the absorption spectra of H_2O, CO, CO_2, CH_4, NH_3, HCl, HBr, and CH_3OH. To make use of these calibration points it is again necessary to refer to the spectrograms included in the paper. In certain regions where the calibration points are far apart the authors use a formula suggested by McKinney and Friedel (1948)

$$T + T_0 = B/(\bar{\nu}_0 - \bar{\nu})^2$$

where T is the instrumental reading for a wavenumber $\bar{\nu}$, and T_0, B and $\bar{\nu}_0$ are constants.

The calibration of an instrument with a caesium iodide prism is dealt with by Mills, Scherer, Crawford and Youngquist (1955), who recommend the lines of the pure rotation spectra of H_2O, HCl and NH_3 in the 24–54 μm range.

Interference fringes (Edser Butler or Fabry-Perot fringes) may also be used for wavelength calibration. If a pair of plates separated by an air film is placed against the entrance slit, there is interference between the beams passing straight through and those reflected back and forth in the air layer ; if t is the thickness of the air film, maxima of intensity occur at wavelengths given by $2t/n$, $2t/(n+1)$, $2t/(n+2)$, etc., where n is integral and can be determined by comparing the fringes with two known spectral lines in the range (see standard textbooks). The possibility of using such fringes (Plate 2A) as a means of calibration in the infra-red has been investigated, but accuracy may be affected by displacement of the maxima because of finite width and height of slit. The method is dealt with by Brodersen (1956) who shows that it is possible to calibrate a monochromator in the 400–4500 cm^{-1} range with an accuracy of 0·2 cm^{-1}, provided the slit width is kept less than 7 cm^{-1} ; he determines n in the above expression by comparing the fringes with the well-known bands of HCl at 2900 cm^{-1}. Rank and Bennett (1955) point out that if the plates are coated with multiple dielectric films to increase their reflection, and hence the sharpness of the fringes, difficulties may be created by the phase change on reflection not being constant for various wavelengths ; they show how to use the method in these circumstances.

ABSORBANCE STANDARDS

Useful test standards may be solutions used in the normal specimen cells. On the other hand, if standard ' glasses ' are used, any errors due to the cells (length, placing, cleanliness, etc.) can be eliminated. Use of both may help to locate the source of any errors. Vandenbelt (1960) points out that the reproducibility of results with ' glass filters . . . is substantially better than with those using solution standards. This supports the conclusion . . . that cell manipulation is a serious source of error in spectro-photometric observations.'

Materials used as absorbance standards need to fulfil certain conditions:

(*a*) Glasses, and chemicals used for preparing standard solutions, must be stable ; unaffected by heat or light or moisture.

(*b*) Chemical substances must be pure.

(*c*) The variation of optical density with temperature must be small, so that small temperature changes will have no appreciable effect.

(*d*) Substances used for checking the photometric scale only should have broad maxima and minima, so that small errors in wavelength setting and spectral purity have little effect. But for checking the overall performance of an instrument, including wavelength setting and spectral purity, check measurements should be made at a narrow maximum or on the steep side of an absorption band.

(*e*) The standard substance should have bands spread over the useful part of the wavelength scale.

Glass of neutral tint, in parallel polished disks, is suitable for work in the visible region. It is possible to obtain specimens of glass for which a standardizing laboratory has determined the optical densities at various wavelengths in the visible spectrum. If two or more glass disks are placed in series, the total optical density is the sum of the separate values, so that a set of, say, three glasses of optical density 0·4, 0·8 and 1·6, may be used to check a scale of absorbance up to 2·8, which is sufficiently high for most spectrophotometers. It should be noted, however, that if a large number of disks is used in series the resulting absorbance is rather less than the sum of the separate values owing to the effect of inter-reflections. The difference is not appreciable until one has, say, four glasses of absorbance 0·5, when the error may amount to 0·002 ; for eight glasses it may be 0·007. These values are calculated on the supposition that beams reflected between the surfaces of the specimen and the optical surfaces of the instrument may be neglected ; when this reflection becomes appreciable it may affect the degree of error.

In the ultra-violet, a piece of polished Wood's glass may be used ; its thickness is best chosen to give a suitable optical density near the absorption minimum at 3700 Å.

Blackened gauze forms useful standards that have optical densities approximately independent of wavelength and these are sometimes used in calibrating photographic plates. There may, however, be small variations of optical density with changes in wavelength, caused either by imperfections in the black or by diffraction effects, the extent of the latter depending on the aperture of the instrument and on the gauge of the gauze. Heidt and Bosley (1953) found that for the wavelength range 2000–8000 Å the optical densities of two different gauzes varied from 0·204 to 0·209 and from 0·588 to 0·600.

A number of chemical substances have been used as standards for the visible and ultra-violet. For none of them has the last word been said on the absolute values of the specific extinction coefficient. Current information on the most commonly used solutions is summarized below.

Potassium nitrate. A number of collaborative tests have been carried out. The mean results for three such tests for the broad absorption maximum at $\lambda = 302$ nm, together with a determination made at the National Physical Laboratory are quoted below. More detailed information on the results is given in the references quoted. Values are given as molecular extinction coefficients, assuming for KNO_3, $M = 101$:

Photoelectric Spectrometry Group (1965)
  (*PSG Bull.* **16**, 443) $7\cdot12_0$
Photoelectric Spectrometry Group (1949)
  (*Edisbury, PSG Bull.* **1**, 10) $7\cdot20_4$
National Physical Laboratory (1965)
  (*PSG Bull.* **16**, 443) $7\cdot04_6$
Vandenbelt, J. M. (1960)
  (*J. opt. Soc. Amer.* **50**, 24) $7\cdot06_4$

Potassium chromate. Vandenbelt and Spurlock (1955) have published a mean value determined by a number of Cary photoelectric instruments for alkaline potassium chromate solution at the peak of its broad and flat absorption band:

$$K_2CrO_4 \text{ in } 0\cdot05\text{-N KOH at } \lambda = 370 \text{ m}\mu, \ \epsilon_{mol} = 4800\text{--}4820$$

Potassium dichromate. Still more extensive collaborative tests have been made with dichromate solution, which is said to be preferable to chromate because the alkaline solution of the latter will damage glass and silica cells. Dichromate has broad maxima and minima in the ultra-violet at which measurements are not greatly affected by slit width. The mean values set out below were determined from two independent collaborative surveys, in each of which measurements were made on more than seventy photo-electric instruments of several different makes. The extinction coefficients may be considered as established to the precision indicated by the differences in value between the two sets.

Wavelength (nm)	$E_{1\,cm}^{1\%}$ of $K_2Cr_2O_7$ in 0·01-N H_2SO_4	
	Ketelaar *et al.* (1955)	Gridgeman (1951)
350	107·3	107·0
313	48·8	48·9
257	144·8	145·6
235	124·6	125·2

Various authors have drawn attention to the variability of the absorption of potassium dichromate. The position has been summarized by Johnson (1967). He reports results which indicate that the nature of the absorption spectrum varies with concentration, so that Beer's law does not hold exactly. This variation is attributed to a reaction

$$3\,Cr_2O_7'' + 2\,H^+ \rightarrow 2\,Cr_3O_{10}'' + H_2O.$$

In 0·01-N H_2SO_4 the effect is small but finite at the concentrations useful with a 1 cm cell. This means that, for the best reproducibility, measurements should be made at a given concentration. The values quoted above were determined at concentrations of approximately 0·005 per cent and 0·01 per cent.

Von Halban (1944) has suggested astraphloxin perchlorate as a suitable substance for a standard in both visible and ultra-violet. It has absorption maxima at 2775, 5100 and 5400 Å, with minima at 2360, 3775 and 5175 Å. Von Halban has made careful measurements of the molecular extinction coefficients at wavelengths between 2175 Å and 5750 Å.

Anthraquinone and salicylaldehyde in ethyl alcohol have been suggested for the ultra-violet and have been measured by Vandenbelt, Forsyth and Garrett (1945). R. J. Taylor (1942) suggests benzene-azo-*p*-cresol, with a maximum at 3250 Å, for the calibration of spectrophotometers that are to be used for estimation of vitamin A. Potassium picrate was measured by von Halban, Kortum, and Szigeti (1936) and by von Halban and Litmanowitsch (1941).

o

Pyrene dissolved in iso-octane has been used in a collaborative test [Photoelectric Spectrometry Group Bulletin (1965)]. It has sharp bands at 272 and 240 nm.

WINDOW MATERIALS

A number of materials that may be used for cell windows are listed in Table 10.1, which also gives some indication of the wavelength range for which they are transparent. The alkali halides must be used with care because they are hygroscopic, but they are not difficult to repolish satisfactorily for many purposes. Silver chloride will darken if exposed to ultra-violet or intense visible radiation but may be protected by a thin coating of silver sulphide (Kremers, 1946).

TABLE 10.1

Material	Useful wavelength range (μm)	
	from	to
Antimony sulphide		8
Barium fluoride	0·135	12
Cadmium telluride (Irtran 6)	1·5	31
Caesium bromide		52
Caesium iodide		55
Calcium fluoride (Irtran 3)	0·12	11·5
Glass (crown)	0·32	3
Lithium fluoride	0·10	8
Magnesium fluoride (Irtran 1)	0·5	9
Mica		5·3
Magnesium oxide (Irtran 5)		9·5
Potassium bromide		38
Potassium chloride		28
Sapphire	0·14	5·5
Silica	{ 0·16 / 40	4
Silver chloride	0·6	16
Sodium chloride		22
Thallium bromide-iodide (KRS5)	0·6	60
Zinc selenide (Irtran 4)	5	22
Zinc sulphide (Irtran 2)	0·7	14·5

Irtran materials, made by Kodak, are hot-pressed polycrystalline products that can be ground and polished.

The wavelength ranges shown are the useful ones (approximate) for a thickness of 1 mm. The limiting wavelengths are not included unless of practical interest; short-wave limits are not given for most of the alkali halides, for example, because these are not used for the ultra-violet.

TABLE 10.2. Lower limiting wavelengths for which solvents are transparent (Å)

Solvent	1 mm	1 cm	2 cm	4 cm
*Cyclo*hexane	1900	1950	1990	2070
Hexane	1870	2010	2050	2090
Carbon tetrachloride	2450	2570	—	2620
Water	1870	1910	1930	1950
Ethyl alcohol	1980	2040	2090	2140
Chloroform	2230	2370	2430	2460
Isopentane	—	1790	—	—

TABLE 10.3. Solvents for the 15–25 μm range

Solvent	Wavelength (μm) 15 16 17 18 19 20 21 22 23 24 25
Acetone	
Acetonitrile	
n-Alkyl chlorides (amyl-octyl)	
Benzene	
Carbon tetrachloride	
Chlorobenzene	
Chloroform	
Citral	
trans-Decaline	
Dibenzyl-ether	
Dichloroethylene (*cis* or *trans*)	
Diethyl-ketone	
Dioxane	
Ethyl acetate	
Ethyl propionate	
Ethylene dichloride	
Ethylidene dichloride	
n-Heptane	
cyclo-Hexane	
n-Hexane	
Methyl acetate	
Methyl chloroform	
Methyl *cyclo*pentane	
Methylene chloride	
Paraffin, liquid	
Pyridine	
Tetrachlorethylene	

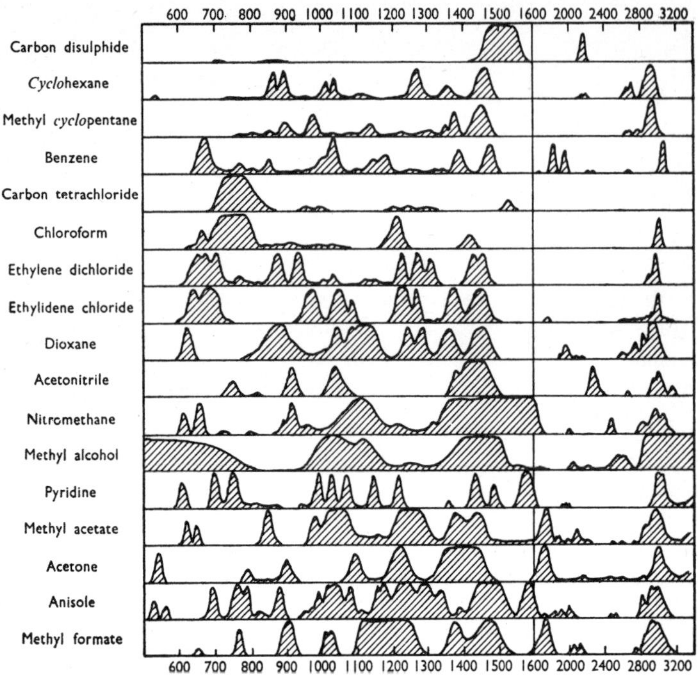

FIG. 10.1. Infra-red absorption bands of a number of solvents. (From Torkington and Thompson, 1945.)

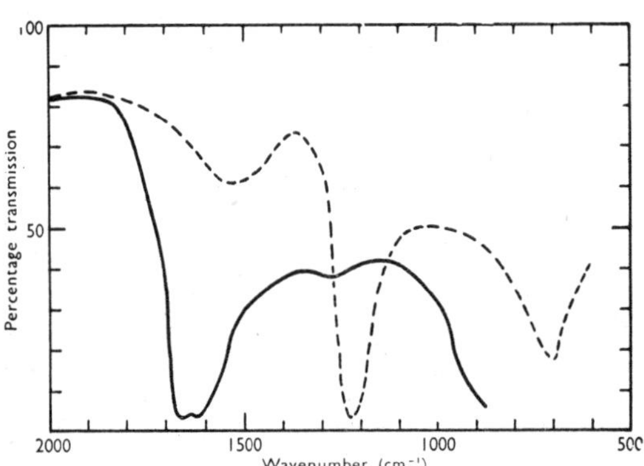

FIG. 10.2. Transmission curves, for a thickness of 0·025 mm, of water and heavy water. (From Blout and Lenormant, 1953.)
H_2O — D_2O - - - -

SOLVENTS

For the visible and ultra-violet

For the earlier work on absorption spectroscopy, the solvents most commonly used were water and alcohol, both of which are polar. Non-polar liquids, such as hexane, heptane or *cyclo*hexane, have since been used to a much greater extent, and carbon tetrachloride and chloroform have also been found to be suitable for certain purposes, although the latter is slightly polar. *Iso*pentane is transparent as far as 1790 Å when used in a 1-cm cell; its purification by treatment with sulphuric acid and washing through silica gel is described by Potts (1952), who also discusses the use of *n*-heptane, 3-methyl-pentane and methyl-*cyclo*hexane in the short-wave ultra-violet. Table 10.2 shows the wavelengths to which some of the solvents are transparent.

All solvents must be specially purified and free of even small traces of absorbing impurities. Some suitable products are available commercially as ' spectroscopic solvents '.

For the infra-red

There is no solvent that is transparent throughout the whole working range up to 30 μm. Carbon tetrachloride and carbon disulphide have the fewest absorption bands. Fig. 10.1 shows the absorption bands of a number of solvents in the 3–17 μm wavelength range. Table 10.3 (compiled from data given by Marrison, 1952) lists a number of solvents for the 15–25 μm range and indicates, by lines, the regions of transparency for a thickness of 0·1 mm. Water has a number of absorption bands in the infra-red. It is however, sufficiently transparent between the bands to be used as a solvent in very thin cells. The absorption curves of thin layers of water and of heavy water (D_2O) are shown in Fig. 10.2. As already mentioned (page 149), the latter may be used to determine absorptions in regions where ordinary water is strongly absorbing.

Further Reading : Bibliography

Further Reading

THE publications listed below are sources of information on absorption spectrophotometry. Only some have been mentioned in the text when discussing particular aspects of the subject. The principal contents are indicated and the following marginal notes are used to point out particularly useful features :

(*a*) Information on theoretical aspects, including the relation of spectra to chemical structure.

(*b*) Helpful details of experimental methods.

(*c*) Absorption data, generally in the form of curves.

ac *Absorption Spectra in the Ultraviolet and Visible Region*, edited by L. Lang (Academic Press, N.Y.). Volumes 1–7 published between 1959 and 1966 include a total of nearly 1200 spectral curves. Volume 1 includes a technical and theoretical introduction.

c *Absorption Spectral Data on Natural Products*, compiled by F. Ellinger (*Tabulae Biologicae*, 1937, **12**, 291). Gives absorption data for a number of natural pigments of biological interest.

a *Advanced Treatise on Physical Chemistry*, by J. R. Partington (Longmans, London, 1954). The first 286 pages of Volume 5 are devoted to the theory of molecular spectra, the relevant quantum theory being developed.

a *Analytical Chemistry*. Review articles are published every two years. The most recent on infra-red and ultra-violet spectroscopy and on spectrofluorimetry appear in a special Review Number published in April 1968. (Vol. 40, pages R116, R246, R255 and R330.)

ab *Analytical Chemistry—The Working Tools*, edited by C. R. N. Strouts, J. H. Gilfillan and H. N. Wilson (O.U.P., London, 1955). This two-volume work, with contributions by members of the staff of Imperial Chemical Industries Ltd., includes in Volume II about 200 pages on absorption spectrophotometry, particularly instrumentation and the relation of spectra to chemical structure.

[a] *Annual Reports on the Progress of Chemistry* (Chemical Society). These include, at fairly regular intervals, sections on various branches of spectroscopy.

Bibliography of Published Information on Infra-red Spectrscopy (H.M. Stationery Office, London). Several volumes have been published at various dates.

[a] *Chemical Applications of Spectroscopy*, edited by W. West (Interscience, New York, 1956—Volume 9 of *Technique of Organic Chemistry*). Comprehensive account of the quantum theory of atomic and molecular spectra and the relationship of spectra to molecular structure. Deals with the ultra-violet, infra-red and microwave regions.

[c] *Infrared Band Handbook*, with supplements, by H. A. Szymanski (–1966, Plenum Press, N.Y.). Tabulates some 18,000 absorption peaks in order of wave number from 36000–100 cm^{-1}. An index of chemical formulae is also included.

[c] *Infra-red Spectra and Characteristic Frequencies of Inorganic Ions*, by F. A. Miller and C. H. Wilkins (*Anal. Chem.*, 1952, **24,** 1253). Absorption spectra of 160 inorganic ions, covering the range 2–16 μ.

[ac] *Infra-red Spectra of Complex Molecules*, by L. J. Bellamy (Methuen, London, 2nd edition, 7th impression, 1966). Includes five charts relating absorption bands to various molecular groups, and about thirty absorption curves covering the range 600–1800 cm^{-1} or 600–3800 cm^{-1}.

[c] *Infra-red Spectral Data.* Compiled by the American Petroleum Institute (Research Project 44). A series of absorption curves of hydrocarbons etc. published periodically by the Carnegie Institute of Technology, Pittsburgh.

[c] *International Critical Tables* (McGraw Hill, New York). Volume V (1929) contains a number of curves showing ultra-violet absorption spectra, and also has a large bibliography.

[ab] *Introduction to Electronic Absorption Spectroscopy in Organic*

Chemistry, by A. E. Gillam and E. S. Stern (Arnold, London, 2nd edition, 1958).

b *Introduction to Quantum Mechanics, with Applications to Chemistry*, by L. Pauling and E. B. Wilson (McGraw Hill, New York, 1935).

c *Landolt-Börnstein Tables* (*Zahlenwerte und Funktionen aus Physik, Chemie, Astronomie, Geophysik und Technik*, Springer, Berlin) includes a number of absorption curves.

ab *Physical Techniques in Biological Research*, by G. Oster and A. W. Pollister (Academic Press, New York, 1955).   Three of the nine chapters are on *Absorption Spectroscopy*, *Ultra-violet Absorption Spectrophotometry* and *Infra-red Spectrophotometry*. These review experimental and theoretical aspects, and include an extensive bibliography.

b *Practical Hints on Absorption Spectrometry*, 200–$800\ m\mu$; by J. R. Edisbury (Hilger & Watts Ltd, London, 1966).   This volume is *not* a systematic textbook, but is a very valuable collection of practical advice based on long experience.

Spectrographic Abstracts (H.M. Stationery Office, London). Published periodically since 1954.   Abstracts the literature on infra-red spectroscopy.

bc *Spectrophotometry in Medicine*, by L. Heilmeyer (Hilger, London, 1943).   Translated by A. Jordan and T. L. Tippell.   An outline of technique, with considerable data on the absorption of body fluids in the visible and ultra-violet.

Ultra-violet and Visible Absorption Spectra, Index for 1934–54, compiled by H. M. Hershenson (Academic Press, New York, 1956).   Gives references to papers that include absorption curves.   There is an index of names of compounds.

c *Ultra-violet Spectra of Aromatic Hydrocarbons*, compiled by R. A. Friedel and M. Orchin (Wiley, New York, 1951).   An atlas of 580 absorption curves, with introduction.

Wavelength Standards in the Infrared, by Rao, K. N., Humphreys, C. J. and Rank, D. H. (1966, Academic Press, N.Y. and London), includes emission and absorption lines suitable for calibration covering the range $10{,}000$–$1\ cm^{-1}$.

CARD SYSTEMS

abcDocumentation of Molecular Spectroscopy. Published in English by Butterworth (London) and in German by Verlag Chemie (Weinheim), in association with the Infra-red Absorption Data Joint Committee (London) and the Institut für Spektrochemie und Angewandte Spektroskopie (Dortmund). Consists of a double series of punched cards coded for chemical groups and band positions: (1) spectrum cards giving an absorption curve for wavelengths from 2·5 to 50 μ, and (2) literature cards each giving an abstract of a publication on spectra or instrumentation.

c *Sadtler Standard Infrared Spectra*, published by Heyden & Son Ltd (London). The collection includes more than 40,000 spectra.

c *Sadtler Standard Ultra Violet Spectra*, published by Heyden & Son Ltd (London). More than 22,000 spectra.

c *UV Atlas of Organic Compounds.* Published by Butterworth (London). In five volumes, including a total of about 1000 compounds, which are indexed alphabetically and by chromophoric groups.

c *Absorption Spectra* (American Society for Testing Materials, Philadelphia). A series of punched cards giving absorption data, for both the ultra-violet and infra-red, for a large number of compounds.

Bibliography and Author Index

The figures in square brackets at the end of an entry refer to pages in this book.

ACQUISTA, N. A., AND PLYLER, E. K., 1952. *Bur.Stand.J.Res., Wash.,* **49,** 13. [205]

ADAMS, G. A., BRADLEY, R. C., AND MCCALLUM, A. B., 1938. *Biochem. J.,* **32,** 646. [37]

AGWU, I. U. AND GLENN, A. L., 1967. *J. Pharm. and Pharmacol.,* **19,** Suppl. 76S. [98]

ALBRECHT, A. C., AND SIMPSON, W. T., 1955. *J. chem. Phys.,* **23,** 1480. [126]

ALDERSON, J. V., ATHERTON, E. AND DERBYSHIRE, A. N., 1961. *J. Soc. Dyers and Colourists,* **77,** 657. [41]

ALKEMADE, C. T. J. *See* Hermann and Alkemade.

ALLSOPP, C. B., 1937. *Proc. roy. Soc.,* **158A,** 167. [105]

ALLSOPP, C. B., AND SZIGETI, B., 1946. *Cancer Research,* **6,** 14. [111]

ALLSOPP, C. B. *See also* Lowry and Allsopp.

AMBROSE, E. J. *See* Elliott, Ambrose and Temple.

ANDREASEN, A. H. M., KREBS, K., DALSGAARD-PEDERSEN, N. A., CAMRADT-PETERSEN, E., AND KJAER, B., 1950. *J.Soc. chem. Ind., Lond.,* **69,** 312. [40]

ANGSTROM, A. J. [33]

ARCHARD, J. F., CLEGG, P. L., AND TAYLOR, A. M., 1952. *Proc. phys. Soc., Lond.,* **65B,** 758. [44]

ARCHBOLD, E., 1957. *J. sci. Instrum.,* **34,** 240. [169]

ARGABRIGHT, P. A. *See* Trusell, Argabright and McKenzie.

ASHTON, G. C. and TOOTILL, J. P. R., 1956. *Analyst,* **81,** 225. [97]

ATHERTON, E., 1955. *J. Soc. Dyers and Colourists,* **71,** 389. [41]

ATHERTON, E. *See* Alderson, Atherton and Derbyshire.

AVERY, D. G. *See* Conn and Avery.

AVERY, D. G., 1952. *Proc. phys. Soc., Lond.,* **65B,** 425. [44]

AVERY, W. H., AND MORRISON, J. R., 1947. *J. appl. Phys.*, **18,** 960. [56]

BAGGER, S. *See* Woldbye and Bagger.

BAGNESS, J. E. *See* Yuen, Bagness and Myles.

BAHR, E. VON. [33]

BAKER, A. W., 1957. *J. Phys. Chem.* **61,** 450. [171]

BAKKER, C. J., 1952. *J. opt. Soc. Amer.*, **42,** 410. [7]

BALY, E. C. C., 1927. *Spectroscopy* (Longman, London). [4, 101]

BANES, F. W., AND EBY, L. T., 1946. *Industr. Engng. Chem. (Anal.),* **18,** 365. [94, 138]

BARCHEWITZ, P., AND HENRY, L., 1954. *J. Phys. Radium,* **15,** 639. [200]

BARER, R., 1950. *Disc. Faraday Soc.,* **9,** 369. [174]

BARER, R., 1955. *Science,* **121,** 709. [149]

BARNES, R. B., GORE, R. C., LIDDELL, U., AND WILLIAMS, Z. VAN, 1944. *Infrared Spectroscopy* (Rheingold, New York). [119, 138]

BARTHAUER, G. L., JONES, F. V., AND METLER, A. V., 1946. *Industr. Engng. Chem. (Anal.),* **18,** 354. [138]

BARUA, R. K. and RAO, M. V. K., 1966. *Analyst,* **91,** 567. [145]

BAUER, G., 1956. *Z. Phys.,* **145,** 279. [62]

BAXTER, B. H. *See* Puttnam, Baxter, Lee and Stott.

BEALE, R. N., 1951. *Photoelect. Spectr. Gr. Bull.,* No. 4, 89. [55]

BEATTIE, J. R., AND CONN, G. K. T., 1955. *Phil. Mag.,* **46,** 989. [44]

BEAVEN, G. H., AND HOLIDAY, E. R., 1952. *Advanc. Protein Chem.,* **7,** 318. [147, 148]

BECKMANN, T. J. *See* Hamilton and Beckmann.

BEER, A., 1852. *Ann. Phys. u. Chem.,* **163** (Dritte Reihe, **86**), 78. [20]

BELL, J. J. *See* Plyler and Bell.

BELLAMY, L. J., 1954. *The Infra-red Spectra of Complex Molecules* (Methuen, London). [119]

BENNETT, H. E. *See* Rank and Bennett.

BLOUT, E. R., AND LENORMANT, H., 1953. *J. opt. Soc. Amer.,* **43,** 1093. [149, 213]

BOAS-TRAUBE, S. *See* Skinner and Boas-Traube.

BOHR, N. [11, 115]

BONNICHSEN, R.   *See* Theorell and Bonnichsen.

BOOTH, R. G., 1940.   *J. Soc. chem. Ind.*, **59**, 181. [58]

BOR, J., 1952.   *Proc. phys. Soc., Lond.*, **65B**, 753. [43]

BOSLEY, D. E.   *See* Heidt and Bosley.

BOUGUER, P. [19]

BOWDEN, F. P., AND MORRIS, S. D. D., 1934.   *Proc. roy. Soc.*, **115B**, 274. [55]

BRADLEY, R. C.   *See* Adams, Bradley and McCallum.

BRAITHWAITE, J. G. N., 1955.   *J. sci. Instrum.*, **32**, 10. [190]

BRATTAIN, W. H., AND BRIGGS, H. B., 1949.   *Phys. Rev.*, **75**, 1705. [190]

BREWSTER, D. [3]

BRIGGS, H. B.   *See* Brattain and Briggs.

BRINKMAN, C. A.   *See* Ketelaar, Fahrenfort, Haas and Brinkman.

BRIX, P., AND HERZBERG, G., 1954.   *Can. J. Phys.*, **32**, 110. [109]

BRODE, W. R., 1949.   *J. opt. Soc. Amer.*, **39**, 1022. [Table B]

BRODERSEN, S., 1954.   *J. opt. Soc. Amer.*, **44**, 22. [29, 31]

BRODERSEN, S., 1956.   *J. opt. Soc. Amer.*, **46**, 255. [206]

BRO-RASMUSSEN, F.   *See* Morton and Bro-Rasmussen.

BURCH, C. R., 1947.   *Proc. phys. Soc., Lond.*, **59**, 47. [174]

BUREAU OF STANDARDS (U.S.), 1949.   *Circular No.* 484. [11]

BYRAM, A. T.   *See* Wilkinson and Byram.

CAMA, H. R.   *See* Morton and Cama.

CAMRADT-PETERSEN, E.   *See* Andreasen, Krebs, Dalsgaard-Pedersen, Camradt-Petersen and Kjaer.

CANNING, R. G., AND DIXON, P., 1955.   *Analyt. Chem.*, **27**, 877. [135]

CANNON, C. G.   *See* Gebbie and Cannon.

CHANCE, B.   *See* Theorell and Chance.

CHAPPEL, F., AND LOTHIAN, G. F., 1951.   *J. appl. Chem.*, **1**, 475. [39]

CHASMAR, R. P.   *See* Smith, Jones and Chasmar.

CHEN, Y. H., 1936.   *Z. physiol. Chem.*, **241**, 129. [145]

CLARK, R. W. G.   *See* Conn, Clark and Parry.

CLEGG, P. L.   *See* Archard, Clegg and Taylor.

COGGLESHALL, N. D. AND SAIER, E. L. 1946.   *J. appl. Phys.* **17**, 450 [34]

COMMONER, R., 1949. *Science*, **110**, 31. [24]

CONN, G. K. T., 1956. *Amer. J. Phys.*, **24**, 451. [62]

CONN, G. K. T. *See also* Beattie and Conn.

CONN, G. K. T. and AVERY, D. G., 1960. *Infrared Methods, Principles and Applications.* (Academic Press) [72]

CONN, G. K. T., CLARK, R. W. G. AND PARRY, C. M. 1969. *J. Sci. Instrum.*, in press. [44]

CONN, G. K. T., AND EATON, G. K., 1954a. *J. opt. Soc. Amer.*, **44**, 546. [44]

CONN, G. K. T., AND EATON, G. K., 1954b. *J. opt. Soc. Amer.*, **44**, 553. [200]

COOPER, B. S., GILLAM, A. E., LOTHIAN, G. F., AND MORTON, R. A., 1942. *Analyst*, **67**, 164. [Table B]

COXON, R. V. *See* Kay and Coxon.

CRAWFORD, B. *See* Downie, Magoon, Purcell and Crawford, *and* Mills, Scherer, Crawford and Youngquist.

CROSSMAN, J. *See* Winkelman and Crossman.

DAGNALL, R. M., SMITH, R. AND WEST, T. S., 1966. *J. chem. Soc.*, A, 1595. [133]

DAGNALL, R. M., SMITH, R. AND WEST, T. S., 1967. *Analyst*, **92**, 20. [133]

DALE, E. B. *See* Nielsen, Thornton and Dale.

DALSGAARD-PEDERSEN, N. A. *See* Andreasen, Krebs, Dalsgaard-Pederson, Camradt-Petersen and Kjaer.

DAVIDSON, H. R., 1962. *Amer. Paint J.*, **46**, 9. [41]

DAVIDSON, H. R. AND HEMMENDINGER, H., 1958. *J. opt. Soc. Amer.*, **48**, 281. [41]

DAWSON, T. L., DERBYSHIRE, A. N. AND LEMIN, D. R., 1965. *J. Soc. Dyers and Colourists*, **81**, 306. [41]

DE BROGLIE, L. [12]

DEELEY, E. M. *See* Walker and Deeley.

DEICHMANN, W., AND SCHAFER, L. J., 1942. *Ind. Eng. Chem.*, Anal. Ed., **14**, 310. [141]

DERBYSHIRE, A. N. *See* Dawson, Derbyshire and Lemin; *and also* Alderson Atherton and Derbyshire.

DE WITT, J. B., AND SULLIVAN, M. X., 1946. *Industr. Engng. Chem. (Anal.)*, **18**, 117. [145]

DIMROTH, K. VON., 1939. *Angew Chem.*, **52**, 545. [104]

DITCHBURN, R. W., JUTSUM, P. J., AND MARR, G. V., 1953. *Proc. roy. Soc.*, **219A**, 89. [22]

DIXON, P. *See* Canning and Dixon.

DOLANCE, A., AND HEALY, P. W., 1945. *Industr. Engng. Chem. (Anal.)*, **17**, 718. [136]

DOWNIE, A. R., MAGOON, M. C., PURCELL, T., AND CRAWFORD, B., 1953. *J. opt. Soc. Amer.*, **43**, 941. [205]

DREYFUS, R. W. D. *See* Rosenwasser and Dreyfus.

DRIFFIELD, V. C. *See* Hurter and Driffield.

DUNCAN, D. R., 1940. *Proc. phys. Soc.*, **52**, 390. [40]

DUNCAN, D. R., 1962. *J. Oil and Col. Chemists Assn.*, **45**, 320. [40]

DUNCAN, D. R., 1965. *Photoelec. Spectr. Group Bull.*, No. 16, 483. [40]

EATON, G. K. *See* Conn and Eaton.

EBY, L. T. *See* Banes and Eby.

EDISBURY, J. R., 1949. *Photoelect. Spectr. Gr. Bull.*, No. 1, 10. [208]

EINSTEIN, A. [10, 72]

EISENBRAND, J. *See* Halban and Eisenbrand.

ELLIOTT, A., 1953. *Nature, Lond.*, **172**, 359. [125]

ELLIOTT, A., AMBROSE, E. J., AND TEMPLE, R., 1948. *J. opt. Soc. Amer.*, **38**, 212. [200]

ELWELL, W. T. AND GIDLEY, J. A. F., 1961. *Atomic Absorption Spectrophotometry* (Pergamon). [135]

ENGSTROM, A. *See* Glick, Engstrom and Malmstrom.

ERRERA, J., GASPART, R., AND SACK, H., 1940. *J. chem. Phys.*, **8**, 63. [122]

ERRERA, J., AND MOLLET, P., 1936. *Nature, Lond.*, **138**, 882. [122]

FAHRENFORT, J., 1961. *Spectrochim. Acta.*, **17**, 698. [45]

FAHRENFORT, J. AND VISSER, W. M., 1962. *Spectrochim. Acta.*, **18**, 1103. [45, 46, 51]

FAHRENFORT, J. *See* Ketelaar, Fahrenfort, Haas and Brinkman.

FARMER, V. C., 1955. *Chem. & Ind.*, **21**, 586. [171]

FARNELL, G. C., 1954. *Proc. phys. Soc., Lond.*, **67B**, 164. [35]

FISHER, R. A. AND YATES, F., 1957. *Statistical Tables for Biological, Agricultural and Medical Research* (Oliver and Boyd, Edinburgh). [97]

P

FOLEY, H. M., 1946. *Phys. Rev.*, **69,** 616. [32, 34]

FOLLETT, D. H., 1935. *Proc. phys. Soc., Lond.*, **47,** 125. [80]

FORD, M. A., AND WILKINSON, G. R., 1954. *J. sci. Instrum.*, **31,** 338. [171]

FORSYTH, J. *See* Vandenbelt, Forsyth and Garrett.

FOX, J., AND MARTIN, A. E., 1940. *Trans. Faraday Soc.*, **36,** 897. [122, 123, 124]

FRASER, R. D. B., 1953. *J. opt. Soc. Amer.*, **43,** 923. [174]

FRASER, R. D. B., 1956. *J. chem. Phys.*, **24,** 89. [125]

FRASER, R. D. B., AND PRICE, W. C., 1952. *Nature, Lond.*, **170,** 490. [125]

FRENCH, C. C. J. *See* Rose and French.

FREEMAN, S. K., 1965. *Interpretive Spectroscopy* (Reinhold). [119]

FRIEDEL, R. A. *See* McKinney and Friedel.

FRODYMA, M. M. AND LIEU, VAN T., 1967. *Anal. Chem.*, **39,** 814. [143]

FRY, D. L., NUSBAUM, R. E., AND RANDALL, H. M., 1946. *J. appl. Phys.*, **17,** 150. [86, 88 et seq., 138]

GABRIEL, A. H., SWAIN, I. R. AND WALLER, W. A., 1965. *J. sci. Instrum.*, **42,** 95. [164]

GARRETT, A. *See* Vandenbelt, Forsyth and Garrett.

GARTON, W. R. S., 1953. *J. sci. Instrum.*, **30,** 119. [156]

GASPART, R. *See* Errera, Gaspart and Sack.

GEBBIE, H. A., AND CANNON, C. G., 1952. *J. opt. Soc. Amer.*, **42,** 277. [190]

GEBBIE, H. A. AND TWISS, R. Q., 1966. *Rep. Progr. Phys.*, **29,** part 2, 729. [184]

GIDLEY, J. A. F. *See* Elwell and Gidley.

GILLAM, A. E., AND HEY, D. H., 1939. *J. chem. Soc.*, 1170. [105]

GILLAM, A. E. AND STERN, E. S., 1960. *Introduction to Electronic Absorption Spectroscopy in Organic Chemistry* (Arnold). [104]

GILLAM, A. E. *See also* Cooper, Gillam, Lothian and Morton.

GILLHAM, E. J., 1956. *J. sci. Instrum.*, **33,** 338. [169]

GILLARD, R. D., 1963. *Analyst*, **88,** 825. [202]

GLENN, A. L., 1963. *J. Pharm. and Pharmacol. Suppl.*, **15,** 123S. [98]

GLENN, A. L., 1967. *Proc. Soc. anal. Chem.*, **4,** 118. [98]

GLENN, A. L. AND WAHBI, A. M., 1969. *Analyst*, in press.

GLENN, A. L. *See* Agwu and Glenn.

GLICK, D., ENGSTROM, A., AND MALMSTROM, B. G., 1951. *Science.* **114,** 253. [24]

GOLAY, M. J. E., 1947. *Rev. sci. Instrum.,* **18,** 357. [170]

GORDON, R. R., AND POWELL, H., 1945. *J. Inst. Petrol.,* **31,** 191. [86 et seq., 138]

GORE, R. C. *See* Barnes, Gore, Liddell and van Williams.

GOULDEN, J. D. S., 1958. *Trans. Far. Soc.,* **54,** 941. [149, 151]

GOULDEN, J. D. S. AND SHERMAN, P., 1962. *J. Dairy Sci.,* **29,** 47. [151]

GRIDGEMAN, N. T., 1951. *Photoelect. Spectr. Gr. Bull.,* No. 4, 67. [209]

GROH, J., AND PAPP, S., 1930. *Z. phys. Chem.,* **149A,** 153. [24]

GROSJEAN, M., *See* Velluz, Legrand and Grosjean.

GUMPRECHT, R. O., AND SLIEPCEVICH, C. M., 1951. *Light Scattering Functions for Spherical Particles* (University of Michigan). [39]

HAAS, C. *See* Ketelaar, Fahrenfort, Haas and Brinkman.

HALBAN, H. VON AND EISENBRAND, J., 1927. *Proc. roy. Soc.,* A **116,** 153. [75]

HALBAN, H. VON, 1944. *Helv. chim. acta,* **27,** 1032. [209]

HALBAN, H. VON, KORTUM, G., AND SZIGETI, B., 1936. *Z. Elektrochem.,* **42,** 628. [209]

HALBAN, H. VON, AND LITMANOWITSCH, M., 1941. *Helv. chim. Acta,* **24,** 44. [209]

HALFORD, R. S. *See* Newman and Halford.

HALVERSON, F., 1947. *Rev. mod. Phys.,* **19,** 87. [124]

HAM, N. S., WALSH, A., AND WILLIS, J. B., 1952. *J. opt. Soc. Amer.,* **42,** 496. [69]

HAMILTON, D. J. AND BECKMANN, T. J., 1966. *Analyst,* **91,** 817. [141]

HAMMOND, V. J., AND PRICE, W. C., 1953. *J. opt. Soc. Amer.,* **43,** 924. [69]

HANSEN, W. N. AND HORTON, J. A., 1964. *Anal. Chem.,* **36,** 783. [47]

HANSEN, W. N., 1965. *Spectrochem. Acta.,* **21,** 815. [47, 49]

HARDY, A. C., AND PINEO, O. W., 1931. *J. opt. Soc. Amer.,* **21,** 502. [174]

HARRISON, G. A., 1957. *Chemical Methods in Clinical Medicine* (Churchill, London). [147]

HARRISON, G. R., 1925. *J. opt. Soc. Amer.*, **11**, 341. [80]

HARTLEY, W. N., 1883. *Sci. Proc. roy. Dublin Soc.*, **3**, 93. [101]

HARTLEY, W. N. [4, 5]

HARTREE, E. F. *See* Keilin and Hartree.

HAUSSER, KUHN *et. al.* 1935. *Z. Phys. Chem.*, **B29**, 363–454. [103]

HAYASHI, Y. *See* Yamamoto, Kumamaru and Hayashi.

HAYWOOD, F. W., AND WOOD, A. A. R., 1957. *Metallurgical Analysis by means of the Spekker Absorptiometer* (Hilger, London). [132]

HEALY, P. W. *See* Dolance and Healy.

HEER, C. V. *See* Rauch and Heer.

HEIDT, L. J., AND BOSLEY, D. E., 1953. *J. opt. Soc. Amer.*, **43**, 760. [208]

HEILBRON, I. M., KAMM, E. D., AND MORTON, R. A., 1927. *Biochem. J.*, **21**, 78. [106]

HEILMEYER, L., 1943. *Spectrophotometry in Medicine*, translated by Jordon and Tippell (Hilger, London). [146, 147]

HEMMENDINGER, H. *See* Davidson and Hemmendinger.

HENRI, V., 1919. *Etudes de Photochemie* (Gauthier-Villars, Paris). [6, 101]

HENRY, L. *See* Barchewitz and Henry.

HERMANN, R. AND ALKEMADE, C. T. J., 1963. *Chemical Analysis by Flame Photometry* (Interscience, N.Y.). [136]

HERSCHER, L. W. *See* Wright and Herscher.

HERZBERG, G. *See also* Brix and Herzberg.

HEY, D. H. *See* Gillam and Hey.

HINTEREGGER, H. E., AND WATANABE, K., 1953. *J. opt. Soc. Amer.*, **43**, 604. [164]

HODSON, A. Z. AND NORRIS, L. C., 1939. *J. biol. Chem.*, **131**, 621. [58]

HOGNESS, T. R., ZSCHEILE, F. P., AND SIDWELL, A. E., 1937. *J. phys. Chem.*, **41**, 379. [67]

HOLIDAY, E. R. *See* Beaven and Holiday.

HOOGE, F. N. *See* Ketelaar and Hooge.

HORTON, J. A. *See* Hansen and Horton.

HUGHES, H. K., 1952. *Analyt. Chem.*, **24**, 1349. [Table B]

HULBERT, E. O., HUTCHINSON, J. F., AND JONES, H. C., 1917. *J. phys. Chem.*, **21**, 150. [22]

HULST, H. C. VAN DE, 1950. *Astrophys. J.*, **112**, 1. [39]

HULST, H. C. VAN DE, 1957. *Light Scattering by Small Particles* (Wiley, New York). [39, 149]

HURTER, F., AND DRIFFIELD, V. C., 1890. *J. Soc. chem. Ind., Lond.*, **9**, 455. [20]

HUTCHINSON, J. F. *See* Hulbert, Hutchinson and Jones.

INGRAM, D. J. E., 1955. *Spectroscopy at Radio and Microwave Frequencies* (Butterworth, London). [114]

INN, E. C. Y., 1955. *Spectrochim. Acta*, **7**, 65. [108, 182]

INN, E. C. Y. *See also* Watanabe, Inn and Zelikoff.

JACQUINOT, P., 1960. *Rep. Progr. Phys.*, **23**, 109. [184]

JANAUER, E., MANGAN, J. AND SMITH, F. R., 1967. *Atomic Absorption News Letter*, **6**, No. 1, 3. [136]

JEFFERIES, J. P. *See* Shaw and Jefferies.

JOHNSON, E. A., 1967. *Photoelec. Spectry. Group Bull.*, **17**, 505. [209]

JOHNSON, F. S., WATANABE, K., AND TOUSEY, R., 1951. *J. opt. Soc. Amer.*, **41**, 702. [164]

JONES, F. E. *See* Smith, Jones and Chasmar.

JONES, F. V. *See* Barthauer, Jones and Metler.

JONES, H. C. *See* Hulbert, Hutchinson and Jones.

JONES, L. H., 1953. *J. chem. Phys.*, **21**, 1892. [125, 126]

JONES, L. H., 1954. *J. chem. Phys.*, **22**, 1135. [125]

JONES, R. N., 1952. *J. Amer. chem. Soc.*, **74**, 2681. [38]

JONES, R. V. *See* Paul and Jones.

JUDD, D. B., AND WYSZECKI, G., 1963. *Colour in Business, Science and Industry.* (Wiley, N.Y.). [40]

JURKIN, V. D., KIRKBRIGHT, G. F. AND WEST, T. S., 1966, *Analyst*, **91**, 89. [133]

JUTSUM, P. J. *See* Ditchburn, Jutsum and Marr.

KAMM, E. D. *See* Heilbron, Kamm and Morton.

KAY, R. H., AND COXON, R. V., 1957. *J. sci. Instrum.*, **34**, 233. [146]

KAYSER, H. *Handbuch der Spektroscopie* (Herzel, Leipzig). [5]

KEILIN, D., AND HARTREE, E. F., 1941. *Nature, Lond.*, **148**, 75. [37]

KEILIN, D., AND HARTREE, E. F., 1949. *Nature, Lond.*, **164,** 254. [55]

KESTEVEN, A. B.   *See* Verel, Saynor and Kesteven.

KETELAAR, J. A. A. , AND HOOGE, F. N., 1955.   *J. chem. Phys.*, **23,** 1549. [120]

KETELAAR, J. A. A., FAHRENFORT, J., HAAS, C., AND BRINKMAN, G. A., 1955.   *Photoelect. Spectr. Gr. Bull.*, No. 8, 176. [209]

KIRKBRIGHT, G. F.   *See* Jurkin, Kirkbright and West.

KIRKBRIGHT, G. F., WEST, T. S., AND WOODWARD, C., 1965.   *Anal. Chem.*, **37,** 137. [133]

KJAER, B.   *See* Andreasen, Krebs, Dalsgaard-Pedersen, Camradt-Petersen and Kjaer.

KLUYVER, J. C., AND MILATZ, J. M. W., 1953. *Physica*, **19,** 401. [138]

KLUYVER, J. C., AND MILATZ, J. M. W., 1954. *Physica*, **20,** 427. [138]

KORTUM, G.   *See* Halban, Kortum and Szigeti.

KREBS, K.   *See* Andreasen, Krebs, Dalsgaard-Pedersen, Camradt-Petersen and Kjaer.

KREMERS, H. C., 1946.   *J. opt. Soc. Amer.*, **36,** 349. [210]

KUBELKA, P., AND MUNK, F., 1931.   *Z.f.tech. Physik*, **12,** 593. [40]

KUBELKA, P., 1948.   *J. opt. Soc. Amer.*, **38,** 448. [40]

KUMAMARU, T.   *See* Yamamoto, Kumamaru and Hayashi.

LAMBERT, J. H. [19]

LARNAUDIE, M., 1952.   *C. R. Acad. Sci., Paris*, **234,** 1440. [130]

LEE, S.   *See* Puttnam, Baxter, Lee and Stott.

LEGRAND, M.   *See* Velluz, Legrand and Grosjean.

LEMIN, D. R.   *See* Dawson, Derbyshire and Lemin.

LENORMANT, H.   *See* Blout and Lenormant.

LEWIS, P. C., AND LOTHIAN, G. F., 1954.   *Brit. J. appl. Phys.* Suppl. **3,** 71. [40]

LEWIS, P. C.   *See also* Lothian and Lewis.

LIDDELL, U.   *See* Barnes, Gore, Liddell and van Williams.

LIEBHAFSKY, H. A.   *See* Pfeiffer and Liebhafsky.

LIEU, VAN T.   *See* Frodyma and Lieu.

LITMANOWITSCH, M.   *See* Halban and Litmanowitsch.

LONGHURST, R. S., 1967.   *Geometrical and Physical Optics* (Longmans). [202]

LOTHIAN, G. F., 1941. *J. sci. Instrum.*, **18**, 200. [57]

LOTHIAN, G. F., 1963. *Analyst*, **88**, 680. [26]

LOTHIAN, G. F., AND LEWIS, P. C., 1956. *Nature, Lond.*, **178**, 1342. [36]

LOTHIAN, G. F. *See also* Chappel and Lothian, *and* Cooper, Gillam, Lothian and Morton, *and* Lewis and Lothian.

LOWAN, A. N., 1949. Nat. Bur. Stand. *Appl. Math. Ser.* 4. [39]

LOWRY, T. M., AND ALLSOPP, C. B., 1937. *Proc. roy. Soc.*, **163A**, 361. [104]

LUFT, K. F., 1943. *Z. tech. Phys.*, **24**, 97. [196]

LUFT, K. F., 1954. *C. R. Acad. Sci., Paris*, **238**, 1651. [198]

LYMAN, T., 1926. *Science*, **64**, 89. [156]

MAGOON, M. C. *See* Downie, Magoon, Purcell and Crawford.

MALMSTROM, B. G. *See* Glick, Engstrom and Malmstrom.

MANGAN, J. *See* Januer, Mangan and Smith.

MARR, G. V. *See* Ditchburn, Jutsum and Marr.

MARRISON, L. W., 1949. *J. Soc. chem. Ind., Lond.*, **68**, 192. [139]

MARRISON, L. W., 1952. *J. sci. Instrum.*, **29**, 233. [211]

MARTIN, A. E., 1952. *Industrial Chemist*, **28**, 243. [83]

MARTIN, A. E. *See also* Fox and Martin.

MASON, S. F., 1962. *Molec. Phys.*, **5**, 343. [128]

MASON, S. F., 1963. *Quartly Revs. chem. Soc.*, **17**, 20. [129]

MAYNEORD, W. V., AND ROE, E. M. F., 1937. *Proc. roy. Soc.*, **158A**, 634. [54]

MCCALLUM, A. B. *See* Adams, Bradley and McCallum.

MCCUBBIN, T. K., AND SINTON, W. M., 1952. *J. opt. Soc. Amer.*, **42**, 113. [183, 184]

MCDERMOTT, L. H. *See* Preston and McDermott.

MCKENZIE, W. F. *See* Trusell, Argabright and McKenzie.

MCKINNEY, D. S., AND FRIEDEL, R. A., 1948. *J. opt. Soc. Amer.*, **38**, 222. [205]

MECKE, R., AND OSWALD, F., 1951. *Z. Phys.*, **130**, 445. [205]

MELHUISH, W. H., 1960. *J. phys. Chem.*, **64**, 762. [194]

METLER, A. V. *See* Barthauer, Jones and Metler.

MIE, G., 1908. *Ann. Phys., Lpz.*, **25**, 377. [39]

MILATZ, J. M. W. *See* Kluyver and Milatz.

MILLER, W. A.,   1862–4. *J. chem. Soc.*, **17**, 59. [3]

MILLER, W. A.   *See also* Stokes and Miller.

MILLIKAN, G. A., 1942. *Rev. sci. Instrum.*, **13**, 434. [145]

MILLS, I. M., SCHERER, J. R., CRAWFORD, B., AND YOUNGQUIST, M., 1955. *J. opt. Soc. Amer.*, **45**, 785. [205]

MILLS, I. M., AND THOMPSON, H. W., 1955. *Proc. roy. Soc.*, **228A**, 287. [54, 130]

MOLLET, P.   *See* Errera and Mollet.

MORRIS, C. J. O. R., 1944. *J. Physiol.*, **102**, 441. [147]

MORRIS, S. D. D.   *See* Bowden and Morris.

MORRISON, J. R.   *See* Avery and Morrison.

MORTON, R. A., 1942.   *The Application of Absorption Spectra to the Study of Vitamins, Hormones and Co-enzymes* (Hilger, London). [92, 101-2, 104, 112, 113, 127, 142, 144]

MORTON, R. A., AND BRO-RASMUSSEN, F., 1955.   *Analyst*, **80**, 410. [143]

MORTON, R. A., AND CAMA, H. R., 1953.   *Analyst*, **78**, 74. [143]

MORTON, R. A., AND STUBBS, A. L., 1946. *Analyst*, **71**, 348. [94, 143]

MORTON, R. A.   *See also* Cooper, Gillam, Lothian and Morton, *and* Heilbron, Kamm and Morton.

MULDER, E. J., SPRUIT, F. J., AND KEUNING, K. J., 1963.   *Pharm. Weekblad.*, **98**, 745. [94]

MULLIKEN, R. S., 1939. *J. chem. Phys.*, **7**, pp. 14, 20, 121. [104, 130]

MUNK, F.   *See* Kubelka and Munk.

MYLES, D.   *See* Yuen, Bagness and Myles.

NEWMAN, R., AND HALFORD, R. S., 1948.   *Rev. sci. Instrum.*, **19**, 270, [200]

NIELD, C. H., RUSSELL, W. C., AND ZIMMERLI, A., 1940.   *J. biol. Chem.*, **136**, 73. [145]

NIELSEN, J. R., AND SMITH, D. C., 1943.   *Industr. Engng. Chem. (Anal.)*, **15**, 609. [85]

NIELSEN, J. R., THORNTON, V., AND DALE, E. B., 1944.   *Rev. mod. Phys.*, **16**, 307. [34]

NORRIS, K. P.   *See* Wilkins and Norris.

NORRIS, L. C.   *See* Hodson and Norris.

NUSBAUM, R. E.   *See* Fry, Nusbaum and Randall.

O'DONNELL, M. J. O.   *See* Stimson and O'Donnell.

OKSENGORN, B., 1955.   *C. R. Acad. Sci., Paris*, **240,** 2300. [34]

OKSENGORN, B., 1956.   *J. Phys. Radium*, **17,** 606. [34]

OSWALD, F., 1954.   *Z. Elektrochem.*, **58,** 345. [52]

OSWALD, F.   *See also* Mecke and Oswald.

PALIK, E. D. AND RAO, K. N., 1956.   *J. chem. Phys.*, **25,** 1174. [120]

PAPP, S.   *See* Groh and Papp.

PARKER, C. A., 1959.   *Analyst*, **84,** 446. [194]

PARKER, C. A. AND REES, W. T., 1958.   *Nature*, **182,** 1002. [193]

PARKER, C. A. AND REES, W. T., 1962.   *Analyst*, **87,** 83. [193]

PARRY, C. M.   *See* Conn, Clark and Parry.

PASCHEN, F., 1897.   *Ann. Phys. u. Chem.*, **296** (Neue Folge **60**), 712. [30]

PAUL, W., AND JONES, R. V., 1951.   *Nature, Lond.*, **168,** 554. [40]

PERKIN-ELMER, 1955.   *Instrument News*, **6** (2), 7. [52]

PERRY, J. W., 1938.   *Proc. phys. Soc., Lond.*, **50,** 265. [62]

PERRY, J. W., 1956.   *Problems in Contemporary Optics*, pp. 265–284 (Istituto Nazionale de Ottica Arcetri-Firenze). [62]

PFEIFFER, H. G., AND LIEBHAFSKY, H. A., 1951.   *J. chem. Educ.*, **28,** 123. [20]

PFUND, A. H., 1927.   *J. opt. Soc. Amer.*, **14,** 337. [160]

PFUND, A. H., 1940.   U.S. Patent 2,212,211. [196]

PHILPOTTS, A. R., THAIN, W., AND TWIGG, G. H., 1947.   *Nature, Lond.*, **159,** 839. [141]

PINEO, O. W.   *See* Hardy and Pineo.

PITTS, E., 1954.   *Proc. phys. Soc. Lond.*, **67B,** 105. [35]

PLANCK, M. [10]

PLYLER, E. K., AND BELL, J. J., 1952.   *J. opt. Soc. Amer.*, **42,** 266. [190]

PLYLER, E. K.   *See also* Acquista and Plyler.

POTTS, W. J., 1952.   *J. chem. Phys.*, **20,** 809. [211]

POWELL, H.   *See* Gordon and Powell.

PRESTON, J. S., AND MCDERMOTT, L. H., 1934.   *Proc. phys. Soc., Lond.*, **46,** 256. [167]

PRICE, W. C., 1955.   *J. Pharm., Lond.*, **7,** 156. [118-9]

PRICE, W. C.   *See also* Fraser and Price *and* Hammond and Price.

PRITCHARD, B. S., 1955.   *J. opt. Soc. Amer.*, **45**, 365. [68]

PURCELL, T.   *See* Downie, Magoon, Purcell and Crawford.

PUTTNAM, N. A., BAXTER, B. H., LEE, S., AND STOTT, P. L., 1966.   *J. Soc. Cosmetic Chemists*, **17**, 9. [51]

RAMSAY, D. A., 1952.   *J. Amer. chem. Soc.*, **74**, 72. [53]

RANDALL, H. M.   *See* Fry, Nusbaum and Randall.

RAND, S. J. AND STRONG, R. L., 1960.   *J. Amer. chem. Soc.*, **82**, 5. [108]

RANK, D. H., AND BENNETT, H. E., 1955.   *J. opt. Soc. Amer.*, **45**, 69. [206]

RAO, K. N.   *See* Palik and Rao.

RAO, M. V. K.   *See* Barua and Rao.

RAUCH, C. J., AND HEER, C. V., 1957.   *Phys. Rev.*, **105**, 914. [131]

REES, W. T.   *See* Parker and Rees.

RICHARDS, P. L., 1964.   *J. opt. Soc. Amer.*, **54**, 1474. [189]

RIMINGTON, C., 1942.   *Brit. med. J.*, **1**, 177. [146]

RITZOW, G. VON, 1934.   *Ann. Phys., Lpz.*, **19**, 769. [157]

ROBINSON, E. J., 1941.   *Amer. J. Physiol.*, **133**, 428. [37]

ROBINSON, T. S., 1952.   *Proc. phys. Soc., Lond.*, **65B**, 910. [44]

ROE, E. M. F.   *See* Mayneord and Roe.

ROGERS, A. R., 1955.   *Analyst*, **80**, 903. [143]

ROLLASON, E.   *See* van Someren and Rollason.

ROSE, H. E., AND FRENCH, C. C. J., 1948.   *J. Soc. chem. Ind., Lond.*, **67**, 283. [35, 39]

ROSENWASSER, H., AND DREYFUS, R. W. D., 1956.   *J. chem. Phys.*, **24**, 184. [170]

RUSSELL, W. C.   *See* Nield, Russell and Zimmerli.

RUTHERFORD, LORD. [11]

SACK, H.   *See* Errera, Gaspart and Sack.

SAIER, E. L.   *See* Coggleshall and Saier.

SAYNOR, R.   *See* Verel, Saynor and Kesteven.

SCHAFER, L. J.   *See* Deichmann and Schafer.

SCHAWLOW, A. L.   *See* Townes and Schawlow.

SCHERER, J. R.   *See* Mills, Scherer, Crawford and Youngquist.

SCHOTTKY. [73]

SCHRÖDINGER, E. [12]

SCHWUTTKE, G., 1953. *Z. angew. Phys.*, **5,** 303. [42]

SCOTT, W. G., AND TAYLOR, R. J., 1956. *Analyst*, **81,** 117. [143]

SEEDS, W. E., AND WILKINS, M. H. F., 1950. *Disc. Faraday Soc.*, **9,** 417. [174]

SHAW, H. C. AND JEFFERIES, J. P., 1953. *Analyst*, **78,** 519. [96]

SHERMAN, P. *See* Goulden and Sherman.

SIDWELL, A. E. *See* Hogness, Zscheile and Sidwell.

SIMON, I., 1951. *J. opt. Soc. Amer.*, **41,** 336. [44]

SIMPSON, W. T. *See* Albrecht and Simpson.

SINTON, W. M. *See* McCubbin and Sinton.

SKINNER, D. G., AND BOAS-TRAUBE, S., 1947. *Symposium on Particle Size Analysis, Trans. Inst. chem. Engrs., Lond.*, Supplement, p. 57. [39]

SLATER, J. C., 1925. *Phys. Rev.*, **25,** 783. [30]

SLIEPCEVITCH, C. M. *See* Gumprecht and Sliepcevitch.

SMILES. [4]

SMITH, D. C. *See* Nielsen and Smith.

SMITH, R. *See* Dagnall, Smith and West.

SMITH, R. A., JONES, F. E. AND CHASMAR, R. P., 1957. *The Detection and Measurement of infra-red Radiation* (Oxford U.P.). [72, 169]

SNATZKE, G., 1967. *Optical Rotatory Dispersion and Circular Dichroism in Organic Chemistry* (Heyden, London). [202]

SOMEREN, E. VAN, AND ROLLASON, E., 1945. *J. Soc. chem. Ind., Lond.*, **64,** 73. [39]

SPURLOCK, C. H. *See* Vandenbelt and Spurlock.

STEWART, J. E., 1955. *Bur. Stand, J. Res., Wash.*, **54,** 41. [38]

STIMSON, M. M., AND O'DONNELL, M. J. O., 1952. *J. Amer. chem. Soc.*, **74,** 1805. [171]

STOKES, G. G., AND MILLER, W. A., 1862. *Phil. Trans.*, **152,** 606 and 870. [3]

STOTT, P. L. *See* Puttnam, Baxter, Lee and Stott.

STOTT, T. D., 1953. *J. sci. Instrum.*, **30,** 120. [146]

STRONG, R. L. *See* Rand and Strong.

STUBBS, A. L.   *See* Morton and Stubbs.

SULLIVAN, M. X.   *See* De Witt and Sullivan.

SZIGETI, B.   *See* Allsopp and Szigeti *and* Halban, Kortum and Szigeti.

TAYLOR, A. M.   *See* Archard, Clegg and Taylor.

TAYLOR, R. J., 1942.   *Analyst*, **67**, 248. [209]

TAYLOR, R. J.   *See also* Scott and Taylor.

TEMPLE, R.   *See* Elliott, Ambrose and Temple.

THAIN, W.   *See* Philpotts, Thain and Twigg.

THEORELL, H., AND BONNICHSEN, R., 1951.   *Acta chem. scand.*, **5**, 1105. [111]

THEORELL, H., AND CHANCE, B., 1951.   *Acta chem. scand.*, **5**, 1127., [109]

THIELE, J. [103]

THOMPSON, H. W., 1944.   *J. chem. Soc.*, 183. [115]

THOMPSON, H. W., AND TORKINGTON, P., 1946.   *Trans. Faraday. Soc.*, **42**, 432. [121]

THOMPSON, H. W.   *See also* Mills and Thompson *and* Torkington and Thompson *and* Whiffen and Thompson.

THORELL, B., 1950.   *Disc. Faraday Soc.*, **9**, 432. [174]

THORNBERG, W., 1955.   *J. opt. Soc. Amer.*, **45**, 740. [174]

THORNTON, V.   *See* Nielsen, Thornton and Dale.

TOOTILL, J. P. R.   *See* Ashton and Tootill.

TORKINGTON, P. AND THOMPSON, H. W., 1945.   *Trans. Faraday Soc.*, **41**, 246. [213]

TORKINGTON, P.   *See also* Thompson and Torkington.

TOUSEY, R.   *See* Johnson, Watanabe and Tousey.

TOWNES, C. H., AND SCHAWLOW, A. L., 1955.   *Microwave Spectroscopy* (McGraw Hill, New York). [114]

TRUSELL, F., ARGABRIGHT, P. A., AND MCKENZIE, W. F., 1967.   *Anal. Chem.*, **39**, 1025. [136]

TWIGG, G. H.   *See* Philpotts, Thain and Twigg.

TWISS, R. Q.   *See* Gebbie and Twiss.

VANDENBELT, J. M., FORSYTH, J., AND GARRETT, A., 1945.   *Industr. Engng. Chem. (Anal.)*, **17**, 235. [209]

VANDENBELT, J. M., AND SPURLOCK, C. H., 1955.   *J. opt. Soc. Amer.*, **45**, 967. [208]

VANDENBELT, J. M., 1960. *J. opt. Soc. Amer.*, **50**, 24. [206, 208]

VAUGHAN, E. J., 1941. *The Use of the Spekker Photoelectric Absorptiometer in Metallurgical Analysis.* (Institute of Chemistry, London.) [132]

VAUGHAN, E. J., 1942. *Further Advances in the Use of the Spekker Photoelectric Absorptiometer in Metallurgical Analysis* (Institute of Chemistry, London). [132]

VELLUZ, L., LEGRAND, M., AND GROSJEAN, M., 1965. *Optical Circular Dichroism* (Academic Press, London). [202]

VEREL, D., SAYNOR, R., AND KESTEVEN, A. B., 1961. *Spectrovision*, No. 10, 1. [146]

VIERORDT, K., 1873. *Die Anwendung des Spektral-Apparates zur Photometrie der Absorptions-Spektren* (Tübingen). [22]

VINCENT-GEISSE, J., 1955. *Ann. Phys., Paris*, **10**, 185. [54]

VISSER, W. M. *See* Fahrenfort and Visser.

WALKER, P. M. B., AND DEELEY, E. M., 1955. *Photoelect. Spectr. Gr. Bull.*, **8**, 192. [174, 181]

WALSH, A., 1951. *Nature. Lond.*, **167**, 810. [68]

WALSH, A., 1952. *J. opt. Soc. Amer.*, **42**, 94. [68, 69]

WALSH, A., 1953. *J. opt. Soc. Amer.*, **43**, 58. [69]

WALSH, A., 1955. *Spectrochim. Acta*, **7**, 108. [135]

WALSH, A. *See also* Ham, Walsh and Willis.

WALSH, A. D., 1947. *Trans. Faraday Soc.*, **43**, 158. [102]

WALSTRA, P., 1965. *Spectrovision*, **13**, 7. [151]

WATANABE, K., INN, E. C. Y., AND ZELIKOFF, M., 1953. *J. chem. Phys.*, **21**, 1026. [182]

WATANABE, K. *See* Johnson, Watanabe and Tousey *and* Hinteregger and Watanabe.

WEISSLER, A., 1945. *Industr. Engng. Chem. (Anal.)*, **17**, 695. [135]

WELLS, A. J. *See* Wilson and Wells.

WEST, T. S., 1966. *Analyst.*, **91**, 69 [132]

WEST, T. S. *See* Dagnall, Smith and West; *and* Kirkbright, West and Woodward; *and* Jurkin, Kirkbright and West.

WHIFFEN, D. H., AND THOMPSON, H. W., 1945. *J. chem. Soc.*, 268. [84]

WHITE, J. U., 1947. *J. opt. Soc. Amer.*, **37**, 713. [191]

WHITE, R. G., 1964. *Progress in Infra-red Spectroscopy* 2, 275. [94]

WILKINS, M. H. F., AND NORRIS, K. P., 1952. *Nature. Lond.*, **170,** 883. [174]

WILKINS, M. H. F. *See also* Seeds and Wilkins.

WILKINSON, G. R. *See* Ford and Wilkinson.

WILKINSON, P. G. AND BYRAM, E. T., 1965. *Appld. Optics*, **4,** 581. [156]

WILLIAMS, Z. VAN. *See* Barnes, Gore, Liddell and Williams.

WILLIS, J. B., 1951. *Aust. J. sci. Res.*, **4A,** 173. [31]

WILLIS, J. B. *See also* Ham, Walsh and Willis.

WILSON, E. B., AND WELLS, A. J., 1946. *J. chem. Phys.*, **14,** 578. [53]

WINKELMAN, J. AND CROSSMAN, J., 1967. *Anal. Chem.*, **39,** 1007. [142]

WOLDBYE, F. [202]

WOLDBYE, F. AND BAGGER, S., 1966. *Acta chem. Scand.*, **20,** 1145. [24]

WOOD, A. A. R. *See* Haywood and Wood.

WOOD, R. W., 1910. *Phil. Mag.*, **20,** 770. [160]

WOOD, R. W., 1937. *Nature, Lond.*, **140,** 723. [160]

WOODWARD, C. *See* Kirkbright, West and Woodward.

WORMSER, E. M., 1953. *J. opt. Soc. Amer.*, **43,** 15. [168]

WRIGHT, N., 1941. *Industr. Engng. Chem. (Anal.)*, **13,** 1. [94]

WRIGHT, N., AND HERSCHER, L. W., 1946. *J. opt. Soc. Amer.*, **36,** 195. [198]

WRIGHT, W. D., 1964. *The Measurement of Colour* (Hilger, London). [41]

WYBOURNE, B. G., 1960. *J. opt. Soc. Amer.*, **50,** 84. [79]

YAMAMOTO, Y., KUMAMARU, T. AND HAYASHI, Y., 1967. *Talanta*, **14,** 611. [141]

YATES, F. *See* Fisher and Yates.

YOUNGQUIST, M. *See* Mills, Scherer, Crawford and Youngquist.

YUEN, S. H., BAGNESS, J. E. AND MYLES, D., 1967. *Analyst*, **92,** 375. [140]

ZELIKOFF, M. *See* Watanabe, Inn and Zelikoff.

ZIMMERLI, A. *See* Nield, Russell and Zimmerli.

ZSCHEILE, F. P. *See* Hogness, Zscheile and Sidwell.

Index

Tables A, B and C are on pages 244–6

Table A: Useful constants and relationships

Planck's constant	$h = 6\cdot62 \times 10^{-27}$ erg. sec.		$= 6\cdot62 \times 10^{-34}$ joule-sec.
Electronic charge	$e = 4\cdot80 \times 10^{-10}$ e.s.u.		$= 1\cdot60 \times 10^{-19}$ coulomb
	$= 1\cdot60 \times 10^{-20}$ e.m.u.		
Electronic mass	$m = 9\cdot1 \times 10^{-28}$ g		$= 9\cdot1 \times 10^{-31}$ kg
Velocity of e.m. radiation	$c = 2\cdot998 \times 10^{10}$ cm/sec		$= 2\cdot998 \times 10^{8}$ m/sec.
Avogadro's number	$N_A = 6\cdot02 \times 10^{23}$ per mol		
Mechanical equivalent of heat	$J = 4\cdot18 \times 10^{7}$ erg/cal		$= 4\cdot18$ joule/cal
Boltzmann's constant	$k = 1\cdot380 \times 10^{-16}$ erg/degree		$= 1\cdot380 \times 10^{-23}$ joule/degree

$$1 \text{ electron-volt} = 1\cdot60 \times 10^{-12} \text{ erg} = 1\cdot60 \times 10^{-19} \text{ joule}$$

$$N_A \text{ electron-volts} = 23\cdot1 \text{ kilo-calorie}$$

$$\lambda \text{ (angstrom)} \times V\text{(volts)} = 12{,}398$$
(see p. 8)

$$\lambda \text{ (angstrom)} \times N_A E\text{(kilo cal. per mol.)} = 2\cdot85 \times 10^{5}$$
(see p. 8)

Table B: Comparative Nomenclature

Term				Definition	
Cooper, Gillam, Lothian and Morton (1942)	Brode (1949) Hughes (1952)	Nat. Bur. Stand. (1949)	Symbol	In words	In symbols
Intensity of incident radiation			I_0		
Intensity of transmitted (or reflected) radiation			I		
Transmission or transmittance	Transmittance	Transmittance	t	Ratio of transmitted to incident radiation	$t = I/I_0$
Optical density (or extinction)	Absorbance	Absorbance	A†	Common log. of reciprocal transmission	$A = \log_{10}(1/t) = \log_{10}(I_0/I)$ or $I = I_0 10^{-d}$
Extinction coefficient		Absorbance index	K	Absorbance for unit path-length	$K = A/l$ or $I = I_0\,10^{-Kl}$
Specific extinction coefficient	Absorptivity	Absorbancy index	a	Absorbance for unit path-length and concn.	$a = A/cl$ or $I = I_0\,10^{-acl}$
Molecular extinction coefficient	Molar absorptivity	Molar absorbancy index	ϵ	Specific extinction coefficient for concentration 1 g mol per l.	$\epsilon = A/cl$
Absorption coefficient‡			μ	Natural logarithm of reciprocal transmission	$\mu l = \log_e I_0/I$ or $I = I_0\,e^{-\mu l}$
Absorption index‡			k	*Absorption coefficient multiplied by wave-length divided by $4\pi n$	$k = \mu\lambda/4\pi n$ or $I = I_0 e^{-4\pi nkl/\lambda}$

*n = refractive index.   †Sometimes E, a or d.   ‡ Not in common use for analytical work.

Table C: Electromagnetic radiation

For explanation, see pages 7–9

Spectral Region		Far infra-red	Near infra-red	Green light	Ultra-violet	Vacuum ultra-violet
Wavelength	Å	10^6	10^5	5000	3000	1000
	μm	100	10	0·5	0·3	0·1
	nm	10^5	10^4	500	300	100
Wavenumber cm^{-1}		100	1000	20 000	33 333	100 000
Energy	ergs	2×10^{-14}	20×10^{-14}	4×10^{-12}	$6{\cdot}6 \times 10^{-12}$	20×10^{-12}
	joules	2×10^{-21}	20×10^{-21}	4×10^{-19}	$6{\cdot}6 \times 10^{-19}$	20×10^{-19}
	kcal/mol	0·29	2·9	57·2	95·3	286
	eV	0·0124	0·124	2·48	4·13	12·4
Temperature	°K	290	2900	58 000	96 666	290 000